P9-CIU-960

High School
Geometry
UNLOCKED

Penguin
Random
House

By Heidi Torres and the Staff of The Princeton Review

The Princeton Review
24 Prime Parkway, Suite 201
Natick, MA 01760
E-mail: editorialsupport@review.com

Published in the United States by Penguin Random
House LLC, New York, and in Canada by Random House
of Canada, a division of Penguin Random House Ltd.,
Toronto.

ISBN: 978-1-101-88221-4
eBook ISBN: 978-1-101-88222-1
ISSN: 2471-3023

Editor: Meave Shelton
Production Artist: Deborah A. Silvestrini
Production Editors: Kathy G. Carter and Liz Rutzel

Printed in the United States of America on partially
recycled paper.

10   9   8   7   6   5   4   3   2   1

EDITORIAL
Rob Franek, Senior VP, Publisher
Casey Cornelius, VP Content Development
Mary Beth Garrick, Director of Production
Selena Coppock, Managing Editor
Meave Shelton, Senior Editor
Colleen Day, Editor
Sarah Litt, Editor
Aaron Riccio, Editor
Orion McBean, Editorial Assistant

RANDOM HOUSE PUBLISHING GROUP
Tom Russell, Publisher
Alison Stoltzfus, Publishing Manager
Jake Eldred, Associate Managing Editor
Ellen L. Reed, Production Manager

Contents

Acknowledgments

The Princeton Review would like to emphatically thank Heidi Torres for her outstanding work and unwavering dedication to this project, Chad Chasteen for his vision and guidance, Maurice Kessler for his peerless attention to detail and design, Gabe Berlin and Keren Peysakh for their illustration wizardry, and Deborah A. Silvestrini for her nonstop, can-do attitude toward laying out this book. Special thanks also to John Yearley, Chris Knuth, Chris Chimera, Sara Kuperstein, Kathy G. Carter, and Liz Rutzel.

Register Your

1 Go to **PrincetonReview.com/cracking**

2 You'll see a welcome page where you can register your book using the following ISBN: 9781101882214

3 After placing this free order, you'll either be asked to log in or to answer a few simple questions in order to set up a new Princeton Review account.

4 Finally, click on the "Student Tools" tab located at the top of the screen. It may take an hour or two for your registration to go through, but after that, you're good to go.

If you are experiencing book problems (potential content errors), please contact EditorialSupport@review.com with the full title of the book, its ISBN number (located above), and the page number of the error. Experiencing technical issues? Please e-mail TPRStudentTech@review.com with the following information:

- your full name
- e-mail address used to register the book
- full book title and ISBN
- your computer OS (Mac or PC) and Internet browser (Firefox, Safari, Chrome, etc.)
- description of technical issue

Book Online!

Once you've registered, you can...

- Access and download "Key Points" review sheets for each chapter

- Work through additional Locksmith sample problems.

- Delve deeper into geometry with a bonus chapter on probability.

- Consult a printable glossary of terms used in the book to make sure you've got everything straight.

- Download full-size versions of some of the geometry figures in this book

Look For These Icons Throughout The Book

 Online Supplements

 Online Practice

About This Book

WHY HIGH SCHOOL UNLOCKED?

It might not always seem that way, especially after a night of endless homework assignments, but high school can fly by. Classes are generally a little larger, subjects are more complex, and not every student has had the same background for each subject. Teachers don't always have the time to re-explain a topic, and worse, sometimes students don't realize that there's a subject they don't fully understand. This feeling of frustration is a bit like getting to your locker and realizing that you've forgotten a part of the combination to open it, only there's no math superintendent you can call to clip the lock open.

That's why we at The Princeton Review, the leaders in test prep, have built the *High School Unlocked* series. We can't guarantee that you won't forget something along the way—nobody can—but we can set the tools for unlocking problems at your fingertips. That's because this book not only covers all the basics of geometry, but it also focuses on alternative approaches and emphasizes how all of these techniques connect with one another.

How to Use This Book

The speed at which you go through this material depends on your personal needs. If you're using this book to supplement your daily high-school classes, we recommend that you stay at the pace of your class, and make a point out of solving problems in both this book and your homework in as many ways as you can. This is the most direct way to identify effective (and practical) tools.

If, on the other hand, you're using this book to review topics, then you should begin by carefully reviewing the Goals listed at the start of each chapter, and taking note of anything that seems unfamiliar or difficult. Try answering some of the example problems on your own, as you might just be a little rusty. Applying math skills is a lot like riding a bike, in that it comes back quickly—but that's only true if you learned how in the first place. Take as much time as you need, then, to connect with this material. As a real test of your understanding, try teaching one of these troublesome topics to someone else.

Ultimately, there's no "wrong" way to use this book. You wouldn't have picked this up if you weren't genuinely interested, so the real key is that you remain patient and give yourself as much time as you need before moving on. To aid in this, we've carefully designed each chapter to break down each concept in a series of consistent and helpful ways.

Goals and Reflect

Each chapter begins with a clear and specific list of objectives that you should feel comfortable with by the end of the chapter. This allows you not only to assess which sections of the book you need to focus on, but also to clarify the underlying skills that each example is helping to demonstrate. Think of this sort of goal-based structure as a scavenger hunt: It's generally more efficient to find something if you know what you're looking for.

Along those lines, each chapter ends with an opportunity to self-assess. There's no teacher to satisfy here, no grade to be earned. That said, you're only hurting yourself if you skip past something. Would you really want to jump into the pool before learning how to swim? You've got to determine whether you feel comfortable enough with the foundational understanding of one chapter before diving deeper.

Review

Review boxes serve as quick reminders of previously learned content—formulas and definitions—that are now being expanded on, or utilized to demonstrate new techniques. You'll see a Review box at the start of a lesson; they're a bit like the lists of ingredients provided for a recipe, in that they ensure that you have everything you need to make (learn) something new.

Examples

Each lesson is hands-on, filled with a wide variety of examples on how to approach each type of problem. You're encouraged to step in and solve the problem yourself at any time, and each step is clearly explained so that you can either compare techniques or establish a successful strategy for your own coursework. Moreover, examples gradually increase in difficulty throughout the chapter. In moving from basic expressions to more complicated real-world examples, we hope to provide a clearer picture of how to break down concepts to their core techniques.

Locksmith Questions

This element features sample questions in the style of the ACT and SAT, providing you with an enhanced opportunity to see how these topics might appear on a high-stakes test. Answers and explanations appear separately on the next set of pages, allowing you to see how the skills and strategies that you're developing can be adapted or focused, regardless of the context in which a question appears.

Key Chains

Our Key Chain sidebars operate much like your actual key chains—they help to sort ideas and, through repetition and connection, to help ensure that you don't misplace them. After all, it doesn't matter how many keys you own if you can't find them when you need them! If you have the time, Key Chains encourage you to go back to earlier questions and lessons with new insights; if not, they double as mini-Review boxes, helping to remind you of all the applicable skills at your disposal.

End-of-Chapter Practice Questions and Solutions

It can be tricky to accurately assess how well you know a subject, especially when it comes to retaining information. We recommend that you wait a day or two between completing a chapter and tackling its corresponding drill so that have a good measure of how well you've absorbed its contents. These questions intentionally scale in difficulty, and in conjunction with the explanations, help you to pinpoint any remaining gaps.

Key Points

For additional support, we've placed printable key points online. These handy tables summarize the major formulas and concepts taught in each chapter, and serve as excellent review material.

Supplies

While using this book, you may at times need the following supplies to fully complete the problems:

- Calculator (graphing or scientific). Keep this with you every time you use the book.

- Ruler/straightedge

- Compass

- Protractor

- Graph paper and scratch paper

- A larger version of the figure or image you are working with

We'll give you a heads-up in the sidebar if any of these supplies will be needed for a particular section. You can download the supplemental, larger images by accessing your student tools.

WHAT NEXT?

Many of these Unlocked techniques can be applied to other subjects. If you're planning on taking the SAT or ACT, we recommend picking up a book of practice questions or tests so that you can keep those keys nice and sharp. If you're moving on to other courses, or higher-level AP classes, remember the connective strategies that most helped you in this book. Learning how to learn is an invaluable skill, and it's up to you to keep applying that knowledge.

Chapter 1
Translation, Reflection, Rotation

GOALS By the end of this chapter, you will be able to:

- Perform translations, reflections, and rotations of figures

- Describe translations, reflections, and rotations of figures

- Write algebraic expressions for translations, reflections, and rotations of figures

- Perform and describe translations and reflections of functions

- Identify congruent and similar figures

- Classify figures as having reflectional symmetry and/or rotational symmetry

Lesson 1.1
Transformations

In everyday language, **transformation** means change. In geometry, transformation refers to changing a figure. You can transform a figure by moving, flipping, turning, or resizing it.

Examples of Transformations

Two figures are **congruent** if they have the same shape and size. If a figure is moved, flipped, or turned, the original and resulting figures are still congruent. These types of transformations are called **rigid motions**—the location or position of the figure may change, but its shape and size are the same. In other words, lengths and angles of the figure remain the same. (Note that resizing is NOT a type of rigid motion.)

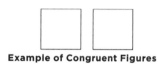

Example of Congruent Figures

When discussing transformations, we refer to the original figure as the **pre-image**, and the new, transformed figure as the **image**. In the figure above, the image and pre-image are congruent.

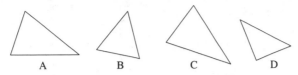

A   B   C   D

Which of the triangles is congruent to triangle A?

(B) (C) (D)

Note that for this question, we're just basing our answer on the way the triangles look. If there were additional information given, such as angle measures and/or side lengths, that would have provided additional support for our answer.

Triangle C is congruent to triangle A. The two triangles have the same shape and size, even though triangle C is rotated slightly.

Triangle B is not congruent to triangle A. Triangle B appears to be the same shape, but it is noticeably smaller than triangle A

Triangle D is not congruent to triangle A. The two triangles have noticeably different shapes.

Two figures are **similar** if they have the same shape, but they may have different sizes. If a figure is proportionally resized, the original and resulting figures are similar, but not congruent.

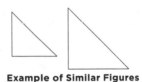

Example of Similar Figures

EXAMPLE 2

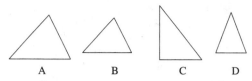

Which of the triangles is similar to triangle A?

(B) (C) (D)

Note that for this question, we're just basing our answer on the way the triangles look. If there were additional information given, such as angle measures and/or side lengths, that would have provided additional support for our answer.

Triangle B is similar to triangle A. The two triangles have the same shape, but different sizes.

Triangle C is not similar to triangle A. The two triangles have noticeably different shapes.

Triangle D is not similar to triangle A. The two triangles have noticeably different shapes.

Lesson 1.2
Translation

Translation refers to moving a figure to a new location. In other words, it's shifting the figure up, down, left, right, or diagonally.

Examples of Translations

The most common way of dealing with translations is in the coordinate plane. You may need to move a figure in a plane, identify how a figure was moved, or write an algebraic expression for a translation.

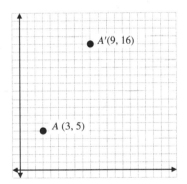

When labeling points, it's customary to use the same letter for the image and pre-image, and usually to use capital letters. You should also add a **prime mark** to the label for the image, to distinguish it from the pre-image. A prime mark is almost like an apostrophe, but straighter, and it's used almost exclusively for mathematics. In the figure above, A is the pre-image and A' is the image.

Take a look at the graph again.

The prime mark is also used to label feet and inches. A single prime mark is used for feet, and a double prime mark for inches. For example, 5'7" means 5 feet and 7 inches.

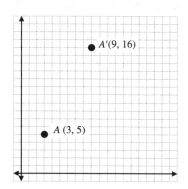

A' is the result of moving A upward and to the right. To find out how much it moved, find the difference between the two coordinates. When looking at the x-coordinates, we see that it moved from 3 to 9. The difference between these coordinates is 6 $(9 - 3 = 6)$; so, this translation moved the point to the right by 6 coordinates. Now do the same for the y-coordinates. The difference between the y-coordinates is 11 $(16 - 5 = 11)$; so, this translation moved the point upward by 11 coordinates.

To describe this translation, you can say that the point has been moved horizontally 6 units, and vertically 11 units. (It makes more sense to describe the x first, since x comes first in x, y coordinate pairs.)

You can also describe the translation in algebraic terms. That is, if the pre-image is (x, y), then the image is $(x + 6, y + 11)$.

Now take a look at this graph.

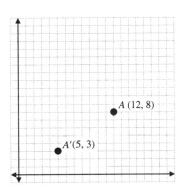

If you're not sure about signs in the coordinate plane, remember this: down is negative, and up is positive; left is negative, and right is positive.

In this example, the point has moved down and to the left. The translation is negative in both directions. We can find the difference between the coordinates, just as we did in the previous example. The difference between the x-coordinates is $(5 - 12)$, or -7. The difference between the y-coordinates is $(3 - 8)$, or -5.

Therefore, this translation is horizontally -7, vertically -5.

We can also express this translation algebraically: $(x - 7, y - 5)$

EXAMPLE 3

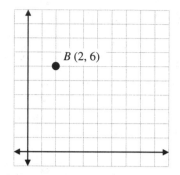

Draw the translated image of *B*, in which the image is moved 4 units in the positive *x* direction and 5 points in the negative *y* direction.

To perform this translation, you're going to move the point right by 4 units, and down by 5 units.

Add 4 to the *x*-coordinate; so, the *x*-coordinate is (2 + 4), or 6.

Subtract 5 from the *y*-coordinate; so, the *y*-coordinate is (6 – 5), or 1.

The coordinate for the image of *B* is (6, 1).

If you prefer, you can count out on the grid to find the point.

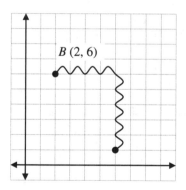

Start at (2, 6). If you count right 4 coordinates, you end up at *x* = 6. Then count down 5, and you're at *y* = 1.

The coordinate for the image of *B* is (6, 1).

Using both methods together is a great way to avoid mistakes!

EXAMPLE 4

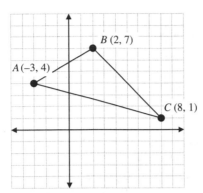

Graph and label the image of triangle *ABC* under the translation (*x* − 2, *y* − 4).

To perform this translation, focus on just the three vertices. Translate those three points, and then use the points to draw the image of the triangle. Move each point left 2, and down 4.

Try it yourself; then read on for the solution.

Coordinate *A* is (−3, 4), so the coordinates of *A′* are $x = (-3 - 2) = -5$, and $y = (4 - 4) = 0$. Point *A′* is at (−5, 0).

Coordinate *B* is (2, 7), so the coordinates of *B′* are $x = (2 - 2) = 0$, and $y = (7 - 4) = 3$. Point *B′* is at (0, 3).

Coordinate *C* is (8, 1), so the coordinates of *C′* are $x = (8 - 2) = 6$, and $y = (1 - 4) = -3$. Point *C′* is at (6, −3).

Your image should have vertices at the following coordinates: *A′* (−5, 0), *B′* (0, 3), and *C′* (6, −3).

Don't forget to draw the triangle!

You can also count out on the grid to find the translated vertices.

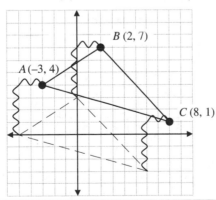

EXAMPLE **5**

Pre-Image	Image
A (0, 0)	A' (–3, 2)
B (5, 0)	B'
C (0, 6)	C'
D (5, 6)	D'

A quadrilateral has the coordinates shown in the table above. If the image has A' at point (–3, 2), what are the coordinates of B', C', and D'?

To find the coordinates for these points, first identify the translation that moves point A to A'. The x-coordinate moves from 0 to –3, so the horizontal move is $x - 3$. The y-coordinate moves from 0 to 2, so the vertical move is $y + 2$. Therefore, we want to translate each coordinate as $(x - 3, y + 2)$. Try it; then check your work below.

The coordinates of B' are $x = (5 - 3) = 2$, and $y = (0 + 2) = 2$. Point B' is at (2, 2).

The coordinates of C' are $x = (0 - 3) = -3$, and $y = (6 + 2) = 8$. Point C' is at (–3, 8).

The coordinates of D' are $x = (5 - 3) = 2$, and $y = (6 + 2) = 8$. Point D' is at (2, 8).

Here is how you may see translations on the ACT.

Rectangle ABCD lies in the standard (x, y) coordinate plane with corners at A (4, 2), B (6, –1), C (1, –4), and D (–1, –1), and is represented by the 2 × 4

matrix $\begin{bmatrix} 4 & 6 & 1 & -1 \\ 2 & -1 & -4 & -1 \end{bmatrix}$. ABCD is then translated, with the corners of the translated

rectangle represented by the matrix $\begin{bmatrix} 1 & 3 & -2 & -4 \\ n & -3 & -6 & -3 \end{bmatrix}$. What is the value of n ?

   A. 0

   B. –1

   C. –2

   D. –3

   E. –4

TRANSLATING FUNCTIONS

If you took Algebra I, you most likely learned a little bit about graphing functions. We'll use basic examples in this section, so for now, don't worry about remembering everything you learned in class. However, you should know that a function takes an input and produces an output. For example, in the function $f(x) = x + 5$, x is the input, and $x + 5$ is the output. So, if $x = 1$, the output is 6 (= 1 + 5), and if $x = 3$, the output is 8 (= 3 + 5).

When working with graphs of functions, remember that input values are graphed on the x-axis, and output values are graphed on the y-axis.

In the examples that follow, you'll see that by changing a function, you can observe fairly predictable changes in the graph of that function.

Consider the function $f(x) = x^2$. In this function, we input a value x and get a value of x^2 as the output. Some examples of points for this function would be $f(2) = 4$, $f(3) = 9$, and so on.

If we graph this function, the graph forms a curve. For every point on this curve, the value of the y-coordinate is the square of the value of the x-coordinate.

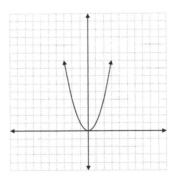

Graph of $f(x) = x^2$

If we change the function slightly, we'll change the appearance of the graph. Let's make a new function to relate to $f(x)$, which we previously defined as $f(x) = x^2$. We'll call the new function $g(x)$, and define it as $g(x) = f(x) + 4$.

What that means is that for every input x, we find the value of $f(x)$ and add 4 to it.

Let's plug in $x = 0$.

$$f(0) = 0$$
$$g(0) = f(0) + 4$$
$$g(0) = 0 + 4$$
$$g(0) = 4$$

This creates the point (0, 4). Some other examples of points in this function are (1, 5), (2, 8), and (–1, 5).

The effect on the graph is that it looks the same as $f(x)$, only it's 4 units higher. It's been translated upward.

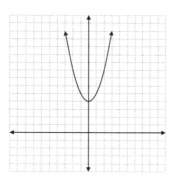

Graph of $g(x) = f(x) + 4$

Compare the points of the original rectangle with the first matrix to see that the x-values of A, B, C, and D run along the top row and their y-values run along the bottom row. For the translated rectangle $ABCD$, plot the points you know: B (3, –3), C (–2, –6), and D (–4, –3). When you've determined those points, note the relationship between them. In your figure, it should be clear that the distance from point C to point D will be the same as the distance from point B to point A. From C to D, the point shifts to the left 2 and up 3. Now do the same thing to point B to get A (1, 0) and $n = 0$, (A). Choices (B), (C), (D), and (E) are in the range of numbers of the problem, but do not translate properly. The correct answer is (A).

Similarly, if we graph a function that equals $f(x) - 4$, we take the graph of $f(x) = x$ and shift it 4 units downward.

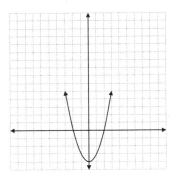

Graph of $g(x) = f(x) - 4$

If you change the expression used for the *input*, the graph will move differently. For example, let's say that $f(x) = x^2$, and a new function $h(x)$ equals $f(x + 4)$. If you plug in a value for x, what you're doing is adding $(x + 4)$ first and *then* inputting that value to $f(x)$.

Let's plug in $x = -4$.

$$h(-4) = f(-4 + 4)$$
$$h(-4) = f(0)$$
$$f(0) = 0$$
$$h(-4) = 0$$

Here is how you may see function translation on the SAT.

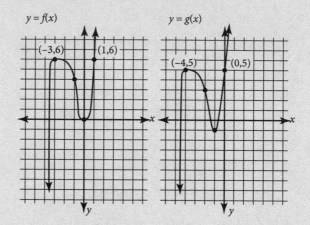

$y = f(x)$

(−3,6) (1,6)

$y = g(x)$

(−4,5) (0,5)

The figures above show the graphs of the functions f and g. The function f is defined by $f(x) = 2x^3 + 5x^2 - x$. The function g is defined by $g(x) = f(x - h) - k$, where h and k are constants. What is the value of hk?

    A. −2

    B. −1

    C.   0

    D.   1

The effect on the graph is that it looks the same as $f(x)$, only it's 4 units to the left. The $h(x)$ function is outputting the $f(x)$ values from 4 units over.

Some other points on this graph are (−3, 1), (−5, 1), (−2, 4), (−6, 4), and so on.

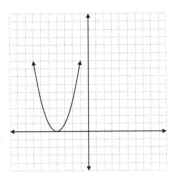

Graph of $h(x) = f(x + 4)$

Similarly, if we graph a function that equals $f(x − 4)$, we take the graph of $f(x) = x$ and shift it 4 units to the right.

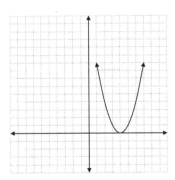

Graph of $h(x) = f(x − 4)$

Thus, the behavior of graphs follows a set of predictable patterns. To make working with graphs easier, try memorizing the following rules for translations:

Translation Rules for Functions

Compared to the graph of $y = f(x)$,

$y = f(x) + C$ moves the graph C units up

$y = f(x) - C$ moves the graph C units down

$y = f(x + C)$ moves the graph C units left

$y = f(x - C)$ moves the graph C units right

The second graph moves down 1 and to the left 1. Remember that when a graph moves to the left, it is represented by $(x + h)$, which would be the same as $x - (-1)$. So $h = -1$. Because a negative k represents moving down, $k = 1$. Therefore, $hk = (-1) \times (1) = -1$, and the correct answer is (B).

Lesson 1.3
Reflection

Reflection is the term for a flipped version of an image. The way we think about reflection in real life, with mirrors, is very much the same way that reflection works in geometry. The reflected image is the same shape and size as the pre-image, but it's backwards.

Examples of Reflections

As with translations, reflections are usually worked in the coordinate plane. Common exercises include drawing a reflection, identifying a reflection line, or writing an algebraic expression for a reflection.

Here is how you may see point reflection on the ACT.

In the standard (x, y) coordinate plane, Z $(-2, -4)$ is reflected over the x-axis. What are the coordinates of the image of Z?

    A. $(4, -2)$

    B. $(4, 2)$

    C. $(2, -4)$

    D. $(-2, 4)$

    E. $(-4, 2)$

One way to create a reflection is to use a folded piece of paper. Draw a figure on one side of the paper; then flip the folded paper over, and trace the image on the other side of the fold. The two figures will be congruent, but flipped. The fold in the paper serves as the **line of reflection**—the line across which the image is flipped.

In an image-editing program, you can use the "flip horizontal" or "flip vertical" feature to perform a reflection.

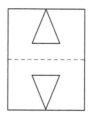

If you draw straight line segments between each pair of reflected points, as shown above, those segments will be parallel to each other. Additionally, the line of reflection forms a **perpendicular bisector** (a line that is perpendicular to another line segment, and intersects at its midpoint) with each of these segments. Each point in the pre-image is the same distance from the reflection line as its mirror image point.

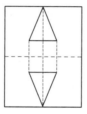

In the image below, point A' is the result of reflecting point A across the y-axis. That is, the reflection line is the y-axis itself.

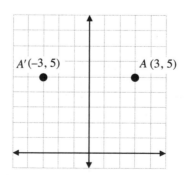

When reflecting a point across an axis, move straight across—that is, perpendicular to—the axis. Make sure the image and pre-image point are the same distance from the axis.

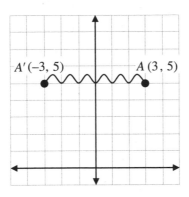

CT A

Draw the standard (x, y) coordinate plane and plot point Z at an x-coordinate of −2 and a y-coordinate of −4. It will look like the image on the left, below.

A reflection of a point over an axis means to use that axis as a sort of mirror line of reflection. If the line of reflection is the x-axis, the image of Z will be above the x-axis. It will still be 2 units from the y-axis and 4 units from the x-axis, like the original point, as shown on the right, below.

The x-coordinate of the image point is still −2, and the y-coordinate is now 4, so (D) is the credited response.

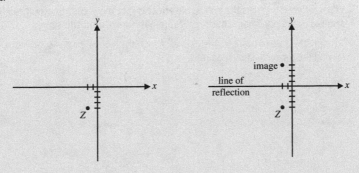

EXAMPLE **6**

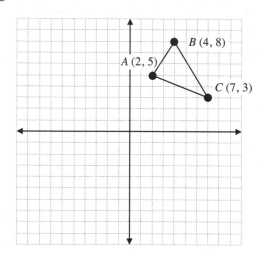

If triangle *ABC* is reflected across the *y*-axis, what are the coordinates of the reflected image's vertices?

Earlier in this chapter, we learned that to translate a polygon, you should just perform the translation on each of its vertices. The same is true for reflection. So, for triangle *ABC*, we're going to perform the reflection on each of its vertices. One way to do this is to use the grid and visually count to where the new vertices will be.

Starting with point *A*, count over to the *y*-axis, which is 2 units from point *A*. So, count over 2 more units, and place point *A'* at coordinate (–2, 5).

With point *B*, count over to the *y*-axis, which is 4 units from point *B*. So, count over 4 more units, and place point *B'* at coordinate (–4, 8).

With point *C*, count over to the *y*-axis, which is 7 units from point *C*. So, count over 7 more units, and place point *C'* at coordinate (–7, 3).

Note that the image and pre-image are congruent, but flipped.

Once you're comfortable with how reflection works, you can memorize this rule for reflection across the *y*-axis:

When reflecting across the *y*-axis, the effect is that the *y*-coordinate(s) will be the same as in the pre-image, but the *x*-coordinates have opposite signs.

To apply this rule to triangle *ABC* above, change the sign of each *x*-coordinate, while leaving the *y*-coordinates the same.

A (2, 5)   *A'* (–2, 5)
B (4, 8)   *B'* (–4, 8)
C (7, 3)   *C'* (–7, 3)

EXAMPLE **7**

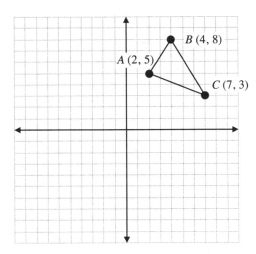

If triangle _ABC_ is reflected across the _x_-axis, what are the coordinates of the reflected image's vertices?

This time, the line of reflection is the _x_-axis. Let's count off on the grid. Starting with point _A_, count down to the _x_-axis, which is 5 units from point _A_. So, count down 5 more units, and place point _A′_ at coordinate (2, –5).

With point _B_, count down to the _x_-axis, which is 8 units from point _B_. So, count down 8 more units, and place point _B′_ at coordinate (4, –8).

With point _C_, count down to the _x_-axis, which is 3 units from point _C_. So, count down 3 more units, and place point _C′_ at coordinate (7, –3).

You can also memorize this rule for reflection across the _x_-axis:

When reflecting across the _x_-axis, the effect is that the _x_-coordinate(s) are the same as in the pre-image, but the _y_-coordinates have opposite signs.

To apply this rule to triangle _ABC_ above, change the sign of each _y_-coordinate, while leaving the _x_-coordinates the same.

_A_ (2, 5)   _A′_ (2, –5)
_B_ (4, 8)   _B′_ (4, –8)
_C_ (7, 3)   _C′_ (7, –3)

Reflection Across the Axis

When reflecting across the y-axis, the effect is that the y-coordinate(s) will be the same as in the pre-image, but the x-coordinates have opposite signs.

When reflecting across the x-axis, the effect is that the x-coordinate(s) are the same as in the pre-image, but the y-coordinates have opposite signs.

EXAMPLE 8

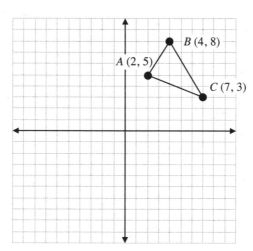

If triangle *ABC* is reflected across the line y = x, what are the coordinates of the reflected image's vertices?

In this example, the line of reflection is the line $y = x$. That line is the set of points for which the x- and y-coordinates are equal—for example, (1, 1), (5, 5), (−5, −5), and so on.

To reflect a coordinate across the line $y = x$, use this simple rule:

When reflecting a point across the line y = x, the effect is that the x-coordinate and y-coordinate are switched.

To apply this rule to triangle *ABC* above, simply switch the x- and y-coordinates of each point.

A (2, 5)	A′ (5, 2)
B (4, 8)	B′ (8, 4)
C (7, 3)	C′ (3, 7)

REFLECTING FUNCTIONS

EXAMPLE 9

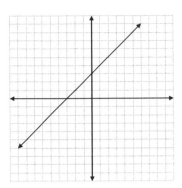

The graph above shows the function $f(x) = x + 3$. What is the result of reflecting this function about the y-axis?

To reflect a function that forms a line, you can choose a few coordinate pairs from the line, and apply the reflection rules to those points. For example, what is $f(2)$ on this function? $f(2) = 2 + 3$, so $f(2) = 5$. That gives us the coordinate pair (2, 5). Great!

To reflect the coordinate (2, 5) across the y-axis, count over, or just change the sign of the x-coordinate. The reflected point is (−2, 5).

We'll need at least one more coordinate to make a line. How about $f(−2)$? $f(−2) = −2 + 3 = 1$. That gives us the coordinate pair (−2, 1). To reflect that coordinate across the y-axis, count over, or just change the sign of the x-coordinate. The reflected point is (2, 1).

Try a few more points for practice! Then, draw a line between your reflected points.

The reflected line looks like this:

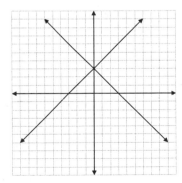

EXAMPLE 10

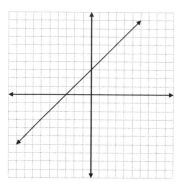

The graph above shows the function $f(x) = x + 3$. What is the result of reflecting this function about the x-axis?

As in the previous example, we'll just choose a few points from the graph, and apply the reflection rules to those points. Let's try $f(3)$, which is $f(3) = 3 + 3$, so $f(3) = 6$. That gives us the coordinate pair (3, 6).

To reflect the coordinate (3, 6) across the x-axis, count downward, or just change the sign of the y-coordinate. The reflected point is (3, –6).

We'll need at least one more coordinate to make a line. Try $f(-4)$. $f(-4) = -4 + 3 = -1$. That gives us the coordinate pair (–4, –1). To reflect that coordinate across the x-axis, count upward, or just change the sign of the y-coordinate. The reflected point is (4, 1).

Try a few more points for practice. Then, draw a line between your reflected points.

The reflected line looks like this:

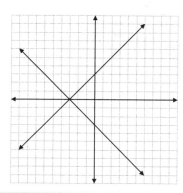

To recap, if you're reflecting across the *x*-axis, you're changing the sign of each of the *y*-coordinates of the figure. If you're reflecting across the *y*-axis, you're changing the sign of each of the *x*-coordinates of your figure.

To see how function reflection is tested on the ACT, access your Student Tools online.

The rules for reflecting across the *x*-axis and *y*-axis can be expressed as follows:

Reflection Rules for Functions

Compared to the graph of $y = f(x)$,

$y = -f(x)$ reflects $f(x)$ across x-axis

$y = f(-x)$ reflects $f(x)$ across y-axis

Lesson 1.4
Rotation

In everyday language, "rotation" means turning or spinning around, such as with a wheel. In geometry, **rotation** means turning a figure around a fixed point. That point is called the **center of rotation**.

You may need to draw a rotation of a figure, identify the angle of rotation, or write an algebraic expression for a rotation.

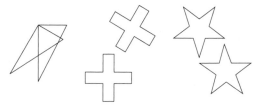

Examples of Rotations

You can imagine a rotation as a pair of hands on a clock. The center of the clock, where both hands are attached, is like the **center of rotation**. The numbers that the hands point to are analogous to the image and pre-image points of a figure.

The figure above illustrates that there is a 60° angle between the 12 and 2. In fact, you can measure the angle from two different directions: clockwise or counterclockwise. (In the example above, the angle measured counterclockwise would be 60°, and the same angle measured clockwise would be –300°.) Both measurements are valid; however, it's important to remember the difference in signs.

> Rotations are **positive** when measured in the counterclockwise direction.
>
> Rotations are **negative** when measured in the clockwise direction.

When you're given a corresponding image and pre-image point, you can measure the **angle of rotation** by drawing the angle between them. More specifically, draw a line segment from the image point to the center of rotation, then another segment from the pre-image point to the center of rotation. The two line segments form an angle, which you can then measure with a protractor.

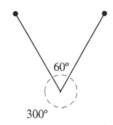

"Clockwise" is the direction of the hands of a clock (i.e. as the hands move from 1 to 2 to 3, and so on). "Counterclockwise" is the opposite direction.

EXAMPLE 11

Supplies

Remember, you can download larger versions of the images in the Example problems from your student tools online.

In the figure above, point _A'_ is the image of _A_, and point _O_ is the center of rotation. Identify the angle of rotation for point _A_ about point _O_.

Here, you're asked to identify the angle of rotation. First, draw a line from point _O_ through point _A_, and another line from point _O_ through point _A'_. Use a protractor to measure the angle between these two segments.

Your measurement should be very close to 45°. (See the illustration below).

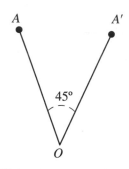

11

EXAMPLE 12

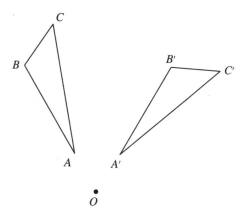

In the figure above, triangle _A'B'C'_ is the image of triangle _ABC_, and point _O_ is the center of rotation. Identify the angle of rotation for triangle _ABC_ about point _O_.

In this example, you're asked to identify the angle of rotation for a triangle. You can do this the same way that you dealt with the single point in Example 11, and in fact, you have to use only one point from the triangle, along with its corresponding image point.

First, choose a point to work with. How about _A_? Draw a line segment from point _O_ through point _A_. Then, draw another line segment from point _O_ through point _A'_. Use a protractor to measure the angle between the two segments.

Your measurement should be very close to 60°. (See the illustration below).

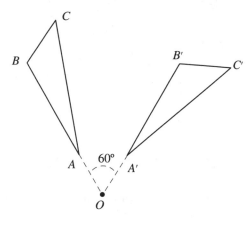

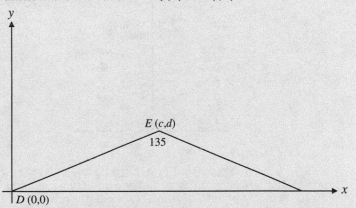

Here is how you may see triangle rotation on the ACT.

Isosceles triangle *DEF* is shown in the standard (*x, y*) coordinate plane below. The coordinates for two of its vertices are *D* (0, 0) and *E* (*c, d*).

Isosceles triangle *DEF* is rotated clockwise by 180° about the origin. At what ordered pair is the image of *E* located?

    A. (−*c, d*)

    B. (*c, −d*)

    C. (−*d ,−c*)

    D. (−*c, −d*)

    E. (*d, c*)

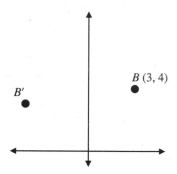

In the figure above, point _B_ has been rotated 90° about the origin, resulting in point _B′_. What are the coordinates of point _B′_?

When dealing with rotations of 90°, one way to approach it is by using rectangles. If you draw a rectangle connecting point _B_ to the _y_-axis and the _x_-axis, as shown below, you can draw the image rectangle by turning it over on its side. The horizontal length of the pre-image becomes the vertical length of the image, and vice versa. In other words, the pre-image rectangle has a horizontal length of 3 units and a vertical length of 4 units; therefore, the image rectangle has a horizontal length of 4 units and a vertical length of 3 units.

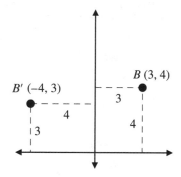

When rotating figures, remember that if you change to a different quadrant in the coordinate plane, you're going to change the sign of one or both coordinates. Check where you are on the axes. In this case, the _x_-coordinate is negative.

Notice how this rotation affected the coordinates: The pre-image coordinate is (3, 4) and the image coordinate is (–4, 3). We'll go over the algebraic rule for this in just a bit.

First, let's see one more example using the rectangle approach.

EXAMPLE 14

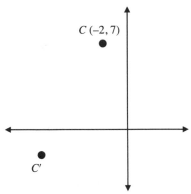

In the figure above, point *C* has been rotated 90° about the origin, resulting in point *C'*. What are the coordinates of point *C'*?

In this example, point *C* has a negative coordinate, but you can approach it the same way as in the previous example. Draw a rectangle connecting point *C* to the *y*-axis and *x*-axis. This rectangle has a horizontal length of 2 and a vertical length of 7. Turn the rectangle on its side (remember to go counterclockwise); it will have a horizontal length of 7 and a vertical length of 2. In this quadrant, both the *x*-coordinate and the *y*-coordinate are negative. The coordinate of *C'* is (–7, –2).

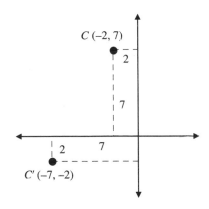

When the triangle is rotated 180°, point *E* will lie in quadrant III. In quadrant III, both coordinates will be negative, which means (A), (B), and (E) can be eliminated. Point *E* will be the same distance from the origin on both the *x*- and *y*-axes, but in the opposite direction, so both coordinates should be the opposite of the original values, or (–*c*, –*d*), as in (D).

You can use the same approach for rotations of 180°, 270°, or 360°, since these angles are multiples of 90°. 180° is two turns of 90°:

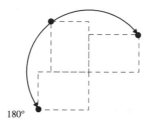

270° is three turns of 90°:

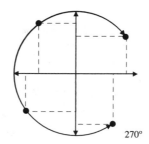

360° is four turns of 90°:

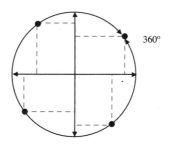

For rotations of 90° and multiples of 90°, use the following rules (note that these rules work only when the center of rotation is the origin):

Rotation Rules for Points

If the center of rotation is the origin (0, 0), then:

$$R_{90°} (x, y) = (-y, x)$$

$$R_{180°} (x, y) = (-x, -y)$$

$$R_{270°} (x, y) = (y, -x)$$

$$R_{360°} (x, y) = (x, y)$$

Lesson 1.5
Symmetry

A figure has **symmetry** if, after certain transformations, the image is identical to the pre-image, and in the same position. There are different kinds of symmetry, including reflectional symmetry and rotational symmetry. Some figures have no symmetry, some have exactly one kind of symmetry, while others may have multiple kinds of symmetry.

REFLECTIONAL SYMMETRY

A figure has **reflectional symmetry** when one half of the image is the mirror image of the other half. In other words, the reflected image just overlaps itself. This is also known as **line symmetry** or **mirror symmetry**. When people talk about "symmetry" in everyday language, they're most likely referring to reflectional symmetry.

Examples of Figures with Reflectional Symmetry

A **line of symmetry** is the line of reflection that results in symmetry.

Think of a line of symmetry as a fold line in a piece of paper. If you can fold a shape and have two identical halves that line up with each other, the shape has line symmetry. A figure can have no lines of symmetry, one line of symmetry, or multiple lines of symmetry.

A rectangle has two lines of symmetry. If you fold it in half vertically or horizontally, you produce two identical halves that line up with each other.

However, if you fold the rectangle in half diagonally, the two halves do not match up, even though they are congruent. This is NOT a line of symmetry.

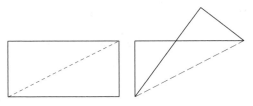

A square has 4 lines of symmetry.

Additionally, any **regular polygon** has the same number of symmetry lines as it does sides. A regular pentagon has 5 lines of symmetry, a regular hexagon has 6 lines of symmetry, and so on.

To see how parabola symmetry is tested on the SAT, access your Student Tools online.

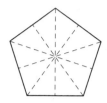

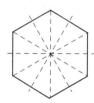

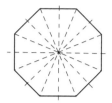

A circle has infinite lines of symmetry!

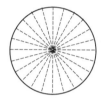

ROTATIONAL SYMMETRY

A figure has **rotational symmetry** if a rotation (other than 0° or 360°) produces the same image overlapping itself.

Examples of Figures with Rotational Symmetry

The **order** of rotational symmetry for a figure is the number of rotations for which the figure has symmetry.

For example, if a figure has order 2 rotational symmetry, then it looks the same at 180°
and 360° rotations.

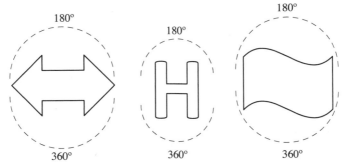

Examples of Figures with Order 2 Rotational Symmetry

There's really no such thing as "order 1" symmetry—that
would just mean that something looks the same when you
turn it completely around, and that's true for every figure.

If a figure has order 3 rotational symmetry, then it looks the same at 120°, 240°, and
360°.

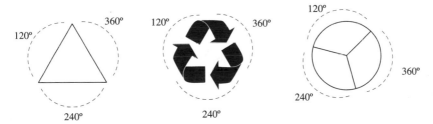

Examples of Figures with Order 3 Rotational Symmetry

A geometric figure has rotational symmetry if it looks the same after a certain
amount of rotation. A geometric figure has reflectional symmetry when one half is
the reflection of the other half. Choice (C) is the only figure that has rotational and
reflectional symmetry.

And so on. The order of symmetry always divides the full rotation evenly.

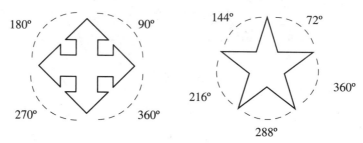

Any **regular polygon** has the same order of symmetry as the number of its sides. For example, a regular pentagon has order 5 symmetry, a regular hexagon has order 6 symmetry, and so on.

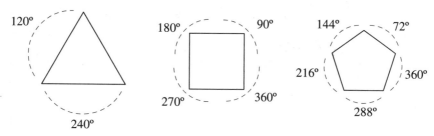

A circle has infinite rotational symmetry!

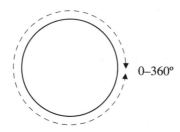

CHAPTER 1 PRACTICE QUESTIONS

Directions: Complete the following problems as specified by each question. For extra practice after answering each question, try using an alternative method to solve the problem or check your work. Larger, printable versions of images are available in your online student tools.

1. *ABCD* is a quadrilateral with coordinates *A* (−5, 2), *B* (6, 1), *C* (1, −5), and *D* (−2, −3). Find the vertices of *A′B′C′D′* under the translation (*x* + 3, *y* − 6).

2. Given *D* (1, 3), *E* (4, 7), *D′* (4, −2), and *E′* (7, 2), determine whether $\overline{DE}$ and $\overline{D′E′}$ are parallel.

3. For each of the coordinates given below, find the new ordered pair by:

    1. Reflecting along the *x*-axis

    2. Reflecting along the *y*-axis

    3. Reflecting along the line *y* = *x*

  a. (5, 5)

  b. (2, −7)

  c. (−3, 0)

  d. (−4, −2)

4. For each of the following questions, check all that apply.

  a. Which of the following types of transformations result in **congruent** figures?

    ☐ Moving a figure (up, down, left, right, or diagonally)

    ☐ Flipping a figure across a line

    ☐ Rotating a figure about a point

    ☐ Changing the size of a figure

  b. Which of the following types of transformations are called **rigid motions**?

    ☐ Moving a figure (up, down, left, right, or diagonally)

    ☐ Flipping a figure across a line

    ☐ Rotating a figure about a point

    ☐ Changing the size of a figure

5. The figure below is the graph of $f(x) = \dfrac{x^2}{2}$.

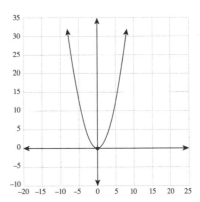

Sketch the graph of $f(x) - 5$.

DRILL

6.

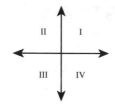

For each of the following, determine which quadrant (I, II, III, or IV) the resulting figure will be in after applying the given reflection:

a.   A circle centered at (−4, 3) with a radius of 1 reflected across the line $x = 0$

b.   A quadrilateral with vertices (−3, −2), (−3, −5), (−7, −5), and (−3, −5) reflected across the line $y = x$

c.   A triangle with vertices located in quadrant IV reflected across the line $y = 1$

d.   A square with vertices (1, 1), (1, 5), (5, 1), and (5, 5) reflected across the line $y = x$

7. Which letters of the alphabet (as written in the style below) have a rotational symmetry of 180°?

A B C D E F G H I J K L M N
O P Q R S T U V W X Y Z

8. The graph below shows a quadrilateral with vertices (5, −5), (10, 10), (15, 10), and (15, 5).

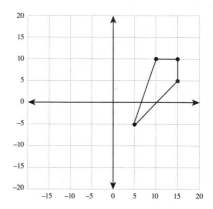

Sketch the rotation of the quadrilateral 270° about the origin.

9. Two dancers are performing a routine, during which they will always maintain precisely 3 feet of space between them. The stage is 30 ft by 20 ft, and the starting position for Dancer A is 10 ft right and 5 ft up, with respect to the bottom left corner (as shown in the figure).

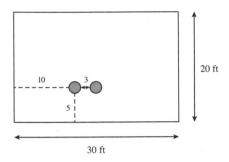

20 ft

30 ft

If the final position of the dancers is represented by a 90° rotation about Dancer A, and a translation of (10, 5) (horizontal and vertical, respectively), what is the final position of the dancers? Sketch your response in the figure (draw the approximate location, and label distances).

Note: All units are in feet.

10. In the figure below, a person has taken a photo of herself while facing a mirror. The image begins to circulate on the Internet, with people debating whether the subject of the photo is holding the phone in her left or right hand.

Only 10% of people get this right!

Comments
(...)
David
OMG people it's a mirror; the phone is in her left hand.
Alex
Seriously? It's definitely the right.

Who is correct here? Is the subject holding her phone in her left or right hand?

SOLUTIONS TO CHAPTER 1 PRACTICE QUESTIONS

1. **The coordinates are $A'(-2, -4)$, B' $(9, -5)$, C' $(4, -11)$, and D' $(1, -9)$.**
    For each coordinate, find A' by adding 3 to each x-coordinate and subtracting 6 from each y coordinate

    a. A $(-5, 2)$ $\rightarrow A'$ $(-5 + 3, 2 - 6)$ $\rightarrow A'$ $(-2, -4)$
    b. B $(6, 1)$ $\rightarrow B'$ $(6 + 3, 1 - 6)$ $\rightarrow B'$ $(9, -5)$
    c. C $(1, -5)$ $\rightarrow C'$ $(1 + 3, -5 - 6)$ $\rightarrow C'$ $(4, -11)$
    d. D $(-2, -3)$ $\rightarrow D'$ $(-2 + 3, -3 - 6)$ $\rightarrow D'$ $(1, -9)$

2. **The segments are parallel.**
    One way to determine whether the segments are parallel is to check the translations of the given coordinates for consistency. That is, see if both pairs of coordinates are translated the same way. Another option, if you know how the slopes of parallel lines are related, is to compare the slopes of the two segments. (You'll learn more about this in Chapter 6). Here's how to compare the translations of the two coordinates:

    D $(1, 3)$ and D' $(4, -2)$: $4 - 1$ is a shift of 3 units to the right and $-2 - 3$ is a shift of -5 units (pay attention to the negative!). This means a shift of 5 units down.

    E $(4, 7)$ and E' $(7, 2)$: $7 - 4$ is a shift of 3 units to the right and $2 - 7$ is a shift of -5 units (pay attention to the negative!). This means a shift of 5 units down.

    Since $\overline{D'E'}$ is created by applying the same shift to each endpoint of $\overline{DE}$, the two lines are parallel to each other.

3. Recall the following rules:

    1. To reflect a coordinate across the x-axis, change the sign of the y-coordinate.
        This is what you should do for step 1 for each coordinate.
    2. To reflect a coordinate across the y-axis, change the sign of the x-coordinate.
        This is what you should do for step 2 for each coordinate.
    3. To reflect a coordinate across the line $y = x$, switch the x- and y-coordinates.
        This is what you should do for step 3 for each coordinate.

    a. Given $(5, 5)$
        1. **$(5, -5)$**   Change the sign of the y-coordinate.
        2. **$(-5, 5)$**   Change the sign of the x-coordinate.
        3. **$(5, 5)$**   Switch the x- and y-coordinates. Since the coordinate already lies on the line $y = x$, the reflected coordinate is the same.

b. Given (2, –7)

    1. **(2, 7)**   Change the sign of the y-coordinate.

    2. **(–2, –7)**   Change the sign of the x-coordinate.

    3. **(–7, 2)**   Switch the x and y coordinates.

c. Given (–3, 0)

    1. **(–3, 0)**   Change the sign of the y-coordinate. Since the coordinate is on the x-axis, the reflected coordinate is the same.

    2. **(3, 0)**   Change the sign of the x-coordinate.

    3. **(0, –3)**   Switch the x- and y-coordinates.

d. Given (–4, –2)

    1. **(–4, 2)**   Change the sign of the y-coordinate.

    2. **(4, –2)**   Change the sign of the x-coordinate.

    3. **(–2, –4)**   Switch the x- and y-coordinates.

4. The answers to both (a) and (b) are the same:

☑ **Moving a figure (up, down, left, right, or diagonally)**
☑ **Flipping a figure across a line**
☑ **Rotating a figure about a point**

If you move, flip, or rotate a figure, the original and resulting figures are congruent. These types of transformations (also known respectively as translations, reflections, and rotations) are called rigid motions.

Congruent figures have the same shape and size. If you change the size of a figure, the original and resulting figures will not be congruent.

5.

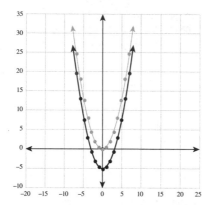

To graph $f(x) - 5$, shift the graph 5 units down. You do not have to plot an exhaustive number of points—just a few to allow you to complete your curve. The vertex (bottom) is a good place to start. It may help to place dots on the points you're trying to move. Choose a point, and then place another dot 5 units directly below it. You may have to eyeball it. In our graph, 5 units is the height of one box.

Note: In this figure, the dots represent the points where $x = 1$, $x = 2$, and so on. This is just for reference; you did not need to figure all that out!

6. a. The center of the circle is located at $(-4, 3)$ and since the radius is 1, the circle is originally located entirely in quadrant II. Reflecting the circle across the line $x = 0$ (the y-axis) flips the new circle into quadrant I.

b. The quadrilateral with the given points is a rectangle located in quadrant III. Reflecting the rectangle across the line $y = x$ requires interchanging the x- and y-coordinates of the original pairs to obtain the new coordinates of $(-2, -3)$, $(-5, -3)$, $(-5, -7)$, and $(-5, -3)$. All four of these new coordinates still remain in quadrant III.

c. Though there are no given vertices for the triangle, it is stated that the triangle is located in quadrant IV. Reflecting the triangle across the line $y = 1$ (a horizontal line parallel to the x-axis passing through $(0, 1)$) moves the triangle into quadrant I.

d. The square with the given vertices is located in quadrant I. The line $y = x$ is one of the lines of symmetry contained in the square, so the rotation produces the exact same square in the exact same location. To confirm, interchange the x- and y-coordinates given to arrive at the new coordinates of $(1, 1)$, $(5, 1)$, $(1, 5)$, and $(5, 5)$. All four of these new coordinates still remain in quadrant I.

7. H, I, N, O, S, X, and **Z.**

A rotational symmetry of 180° means that the figure (or letter, in this case) can be rotated upside down and still be the same as the original. The only letters that have this property are H, I, N, O, S, X, and Z.

8.

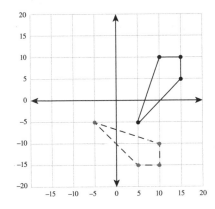

To rotate this figure, perform the rotation on one coordinate at a time. You can use the rectangle method described in the lesson, or the rotation rule $R_{270°} (x, y) = (y, -x)$.

$R_{270°} (-5, 5)$	$\rightarrow (-5, -5)$
$R_{270°} (10, 10)$	$\rightarrow (10, -10)$
$R_{270°} (15, 10)$	$\rightarrow (10, -15)$
$R_{270°} (15, 5)$	$\rightarrow (5, -15)$

9.

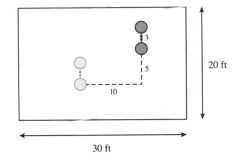

A 90° rotation would result in Dancer B being above Dancer A. The light grey circles in the figure represent the rotation from their original position. Then, the dancers will move 10 feet right and 5 feet up (note the direction for positive number translations).

You don't have to measure this with a ruler...just sketch the location as close as you can get it, and label the distances represented by the translation.

Note: It's also fine if you labeled the distances from a corner of the stage. For example, from the bottom left corner, Dancer A is 20 units right and 10 units up. Don't worry about the size of the circles—consider their location to be measured from their centers.

10. **Alex is correct (right hand).**

    When you stand in front of a mirror, your image is reflected, with the mirror itself acting as the line of reflection. If you hold up your right hand, the reflected image of the right hand will be directly in front of your "real" right hand.

    Some people get confused when they're imagining a person—not a reflection—standing in front of them. If you're facing another person, this is represented by a *rotation*, not a reflection. Your right hand is on their left, and vice versa. But a mirror is not another person!

REFLECT

Congratulations on completing Chapter 1!
Here's what we just covered.
Rate your confidence in your ability to:

- Perform translations, reflections, and rotations of figures
  ① ② ③ ④ ⑤

- Describe translations, reflections, and rotations of figures
  ① ② ③ ④ ⑤

- Write algebraic expressions for translations, reflections, and rotations of figures
  ① ② ③ ④ ⑤

- Perform and describe translations and reflections of functions
  ① ② ③ ④ ⑤

- Identify congruent and similar figures
  ① ② ③ ④ ⑤

- Classify figures as having reflectional symmetry and/or rotational symmetry
  ① ② ③ ④ ⑤

If you rated any of these topics lower than you'd like, consider reviewing the corresponding lesson before moving on, especially if you found yourself unable to correctly answer one of the related end-of-chapter questions.

 Access your online student tools for a handy, printable list of Key Points for this chapter. These can be helpful for retaining what you've learned as you continue to explore these topics.

Chapter 2
Congruence and Theorems

GOALS By the end of this chapter, you will be able to:

- Identify congruent triangles using the SSS, AAS, ASA, SAS postulates

- Use the Pythagorean theorem or the Third Side Rule to solve for an unknown side in a right triangle

- Understand how to write a formal or informal proof

- Construct figures with a compass and straightedge (parallel and perpendicular lines, angle bisector, angle copy, equilateral triangle, square, and regular hexagon)

Lesson 2.1
Triangle Congruence Postulates

POSTULATES: REVIEW THE FOLLOWING POSTULATES, WHICH YOU MAY FIND HELPFUL FOR THIS CHAPTER.

Vertical angles are congruent
(intersection of any two lines)

Alternate exterior angles are congruent
(parallel lines cut by a transversal)

Alternate interior angles are congruent
(parallel lines cut by a transversal)

Corresponding angles are congruent
(parallel lines cut by a transversal)

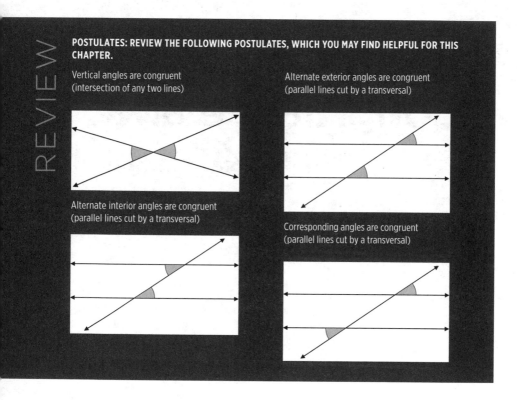

In geometry, we often have to use limited information about a problem in order to figure out several unknowns. Mathematicians call it a **proof** when they use known information to derive other facts and unknowns in a problem. In fact, many geometry classes spend a great deal of time writing exhaustively detailed proofs for problems and figures. We won't make you write full proofs in this book! However, we will show you proofs and sometimes ask you questions about them.

One of the more common facts you'll need to prove is whether or not two figures are congruent. This is one of the most important topics in mathematics, and in fact, a great number of geometric rules and theorems are based on congruence. We deal with this in real life as well—if you're building a structure, you need to make sure that various parts of the structure are congruent (e.g. opposite walls, the ceiling and the floor, etc.) if you don't want the structure to be lopsided. Another example is the tires on your car—if the tires are not exactly congruent (even if one has just a little more or less air in it compared to the others), you'd feel a lot of bumps and shakes as you drive, and it may put excessive wear and tear on your car.

To recap, if two figures are **congruent**, it means that they have the same shape and size. If two polygons are congruent, it means that all of their side lengths and angles are congruent too. Additionally, recall that if a figure is translated, rotated, or reflected, the image and pre-image will be congruent.

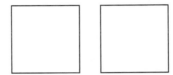

Example of Congruent Figures

Throughout history, mathematicians have developed several rules that we now use to solve problems. The words **theorem** and **postulate** both describe rules. These terms are often used interchangeably, and you probably shouldn't worry about the differences. But just so you know: a postulate (also known as an **axiom**) is something that everyone agrees is true, while a theorem is something that needs to be proven true using logical steps. Many of the "theorems" that we discuss today are effectively postulates, because they've been proven true in the past and we no longer need to doubt their accuracy.

In the case of triangles, we can often derive a lot of information when we know even just a couple of facts (e.g. angle measures or side lengths). In this lesson, you'll learn and understand several basic theorems regarding congruent triangles.

SSS (SIDE-SIDE-SIDE) POSTULATE

EXAMPLE

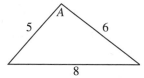 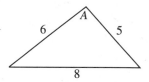

Recall from Lesson 1.1 that if a figure is translated, reflected, or rotated, then the image and pre-image are congruent.

If the triangles above have three side lengths in common, are the triangles necessarily congruent?

If a triangle has three specified side lengths, then both its size and shape are fixed. Those three sides can meet only at specified angles. Any way of "rearranging" the given sides would result in a translation, reflection, or rotation of the same triangle.

For instance, what if we try making angle *A* smaller?

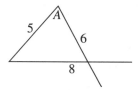 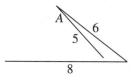

If we make angle *A* smaller, the figure will not be able to connect at the same vertices. We can try moving it around in any way we can think of, but the *only* way to make a triangle with those specific side lengths is if it has the same angles as the original triangle.

Try it yourself! You can demonstrate this postulate using physical objects, such as straws, to form the sides of the triangle.

That's the basic idea behind each of these postulates—once you know certain facts about a triangle, you know that there's only one form that the triangle can take.

SSS (Side-Side-Side) Postulate

If the three sides of one triangle are congruent to the three sides of another triangle, then the triangles are congruent.

SSS Postulate Exercise

For each of the pairs of triangles below, write "yes" if the triangles can be proved to be congruent to the information provided; otherwise, write "no."

Complete the exercise on your own. Answers are at the end of the chapter.

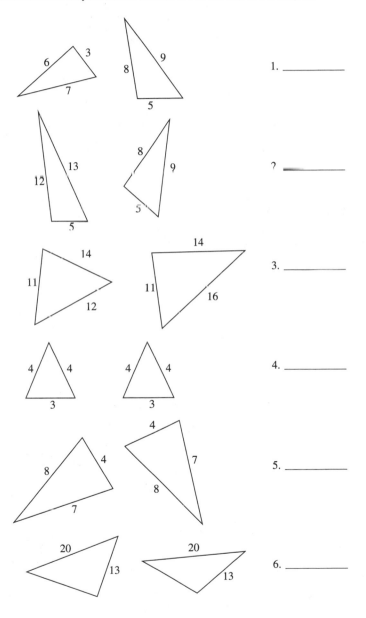

1. _____

? _____

3. _____

4. _____

5. _____

6. _____

AAS (ANGLE-ANGLE-SIDE) AND ASA (ANGLE-SIDE-ANGLE) POSTULATES

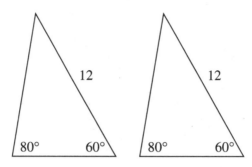

If the triangles above have two angles and one side length in common, are the triangles necessarily congruent?

First, let's check our unknowns in the example above. If two angles are specified, then really, the third angle is specified also. That's because all triangles have angles that add up to 180° (so we can just subtract the two given angles from 180° to find the third one).

If all three angles of a triangle are known, then the *shape* of the triangle is fixed, but not its *size*. You can make the triangle bigger or smaller, while keeping the angles the same.

AAA (Angle-Angle-Angle) *Similarity* Postulate

The AAA (Angle-Angle-Angle) postulate states that if two triangles have all three angle measures in common, then the two triangles are *similar*. However, they are *not* necessarily congruent. Hence, this is often referred to as the "AAA Similarity" postulate.

If all three angles are known, then we know the triangle's shape. If we also specify a side length, then we know the triangle's size.

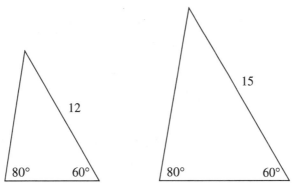

Therefore, we have two congruence postulates that we can use when triangles have two (actually three!) angle measures and one side length in common.

ASA (Angle-Side-Angle) Postulate

If two angles and the included side of one triangle are congruent to the corresponding parts in another triangle, then the triangles are congruent.

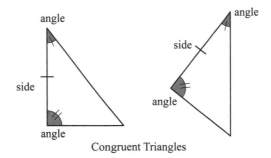

Congruent Triangles

Congruent angles are often identified with marks like these. Look for matching marks (such as a single hash) to identify corresponding angles.

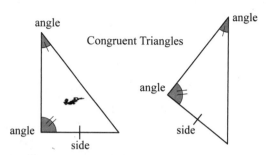

There's an important catch that you must remember—the known side must be in the same place, relative to the angles, in both triangles. That is, if the known side is opposite to the smallest angle in the first figure, then that must also be true for the second figure.

Another way to think of this is that order matters—as you go around the triangle in either direction, you have an angle, side, and angle (or an angle, angle, and side), one after the other, in that order. If the second triangle has those same parts with the same measures in the same order, then the two triangles are congruent.

That's the reason that there are two different postulates specified as AAS and ASA—so we don't mix up the rules and we make sure that the given parts are in the right place.

AAS Postulate Exercise

For each of the pairs of triangles below, write "yes" if the triangles can be proved to be congruent to the information provided; otherwise, write "no."

Complete the exercise on your own. Answers are at the end of the chapter.

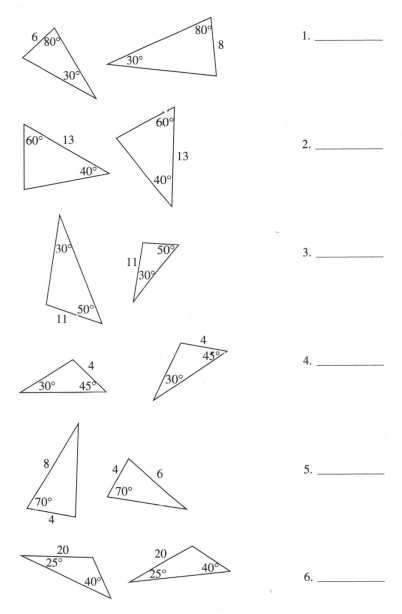

1. _____

2. _____

3. _____

4. _____

5. _____

6. _____

SAS (SIDE-ANGLE-SIDE) POSTULATE

EXAMPLE 3

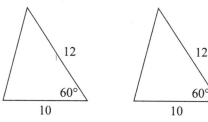

If the triangles above have two sides and one angle in common, are the triangles necessarily congruent?

If a triangle had only two side lengths specified, and no other information, you wouldn't be able to determine the triangle's shape or size. The two given side lengths could come together in any number of ways.

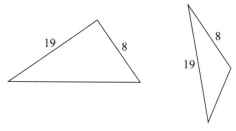

However, if you can specify the angle that joins the two given side lengths, then the triangle is fixed. That angle measure tells you how far apart the other two vertices are, and it specifies the length of the third side of the triangle. (You'll learn how to calculate that length in Chapter 4!)

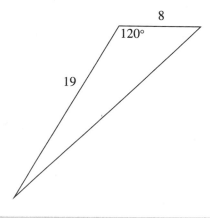

Therefore, the SAS postulate can be used when you have two triangles with two side lengths in common, as well as the angle that joins those two sides.

SAS (Side-Angle-Side) Postulate

If two sides and the included angle of one triangle are congruent to the corresponding parts of another triangle, then the triangles are congruent.

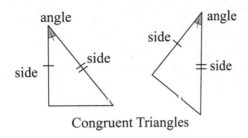

Congruent Triangles

Again, order matters with this postulate, as it does with the ASA and AAS postulates. As you go around the triangle in either direction, you have a side, angle, and side, one after the other, in that order. If the second triangle has those same parts with the same measures in the same order, then the two triangles are congruent.

Additionally, you may wonder if there is an SSA (Side-Side-Angle) postulate. This particular combination does not work for triangle congruence, so *there is **no** SSA postulate.* The reason is that if you have two side lengths and a non-included angle specified, then there are exactly two triangles that can exist with that combination.

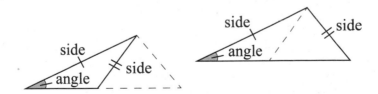

In the example above, there are two different ways the side with two hash marks can fit into the space created by the marked angle.

OTHER IMPORTANT THEOREMS AND POSTULATES FOR TRIANGLES

The following theorems are used very frequently in geometry. You may know them already!

Pythagorean Theorem

The Pythagorean theorem is definitely one of the most important triangle theorems that you should know. It is used to solve for an unknown side length in a **right triangle**.

Pythagorean Theorem

In any **right** triangle with legs a and b and hypotenuse c:

$$a^2 + b^2 = c^2$$

EXAMPLE 4

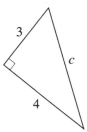

In the triangle above, what is the value of c?

To solve, use the Pythagorean theorem. Plug in the two given side lengths, and solve for the unknown side.

In this case, the given lengths are for the legs of the triangle, which are a and b in the equation. The unknown side is the hypotenuse, or c in the equation.

Plug in:

$$3^2 + 4^2 = c^2$$
$$9 + 16 = c^2$$
$$25 = c^2$$
$$5 = c$$

Take the square root of both sides of the equation.

The value of *c* is 5.

You might have recognized this example as a **Pythagorean triple**—one of the combinations of integers that satisfy the Pythagorean equation. Try memorizing the following Pythagorean triples to make right triangle problems easier. There are many more, but these are some of the most common ones (and the lowest values) that you'll encounter:

$3^2 + 4^2 = 5^2$	$5^2 + 12^2 = 13^2$	$7^2 + 24^2 = 25^2$
$6^2 + 8^2 = 10^2$	$10^2 + 24^2 = 26^2$	$14^2 + 40^2 = 50^2$

Third Side Rule

Use the Third Side Rule if you know two side lengths, but no other information. This will give you a limited range of values for the third, unknown side. It's not a single, exact answer, but sometimes a range is all we have!

To see how the Third Side Rule is tested on the ACT, access your Student Tools online.

The Third Side Rule is also known as the **triangle inequality rule**.

Third Side Rule for Triangles

In any triangle, the length of one side must be greater than the difference and less than the sum of the other two sides.

In other words, find the difference of the two known sides, and also find the sum of the two known sides. The length of the third side has to be *between* these two values.

EXAMPLE 5

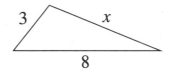

In the triangle above, what is the range of values for _x_?

Use the Third Side Rule to solve for _x_.

First, find the difference of the two known sides:

$$8 - 3 = 5$$

Then, find the sum of the two known sides:

$$8 + 3 = 11$$

Therefore, the value of _x_ must be between 5 and 11.

$$5 < x < 11$$

Here is how you may see the Pythagorean theorem on the ACT.

Greg is making a triangular sail for a boat, shaped like a right triangle and shown below.

To determine how much trim to buy for the sail, Greg calculated the sail's perimeter. What is the sail's perimeter?

A. 275
B. 300
C. 290
D. 220
E. 170

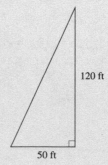

Lesson 2.2
Putting Theorems to the Test

In this lesson, we'll be discussing **proofs**—the logical steps that one can use to show that something is true. In math, that "something" may be a theorem, postulate, definition, or some other property of a figure or problem. As mentioned earlier in this chapter, most geometry courses require students to write many, many complete proofs for any topic you might think of. The reason that this is such a popular teaching topic is that it helps students gain an intuitive understanding of why things work the way that they do. After all, math (or any other subject) shouldn't just be about memorizing facts and formulas. If you don't understand the hows and whys behind your subject, then you're going to struggle when the content gets hard.

This lesson is intended in part to introduce you to the idea of proofs and how they work. We won't actually ask you to write complete proofs, as you would in school. We will, however, show you a few examples of formal proofs as well as informal ones. The more important learning to take away from this lesson is with the concepts behind the proofs—in this case, concepts surrounding triangle congruence.

In this first example, we'll develop not a formal proof, but an explanation for a known postulate.

EXAMPLE

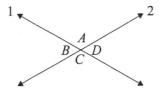

Given that 1 and 2 are lines, explain why $\angle A = \angle C$.

The real question here is why vertical angles are always congruent. You have probably learned the postulate that states that "vertical angles are always congruent," but since we're trying to explain the concept, we're not just going to cite the postulate and call it done. Instead, we'll find reasons to support our explanation.

First, let's consider everything we know about this figure.

Of course, we know that both 1 and 2 are straight lines, which is a fact given in the problem. That's important, because we know that straight lines are 180°. That means that each pair of adjacent angles in the figure has a sum of 180°. When two angles form a line as in these examples, they are called a **linear pair**.

Therefore,

$$\angle A + \angle B = 180°$$
$$\angle B + \angle C = 180°$$
$$\angle C + \angle D = 180°$$
$$\angle D + \angle A = 180°$$

If you look at the list of pairs above, you'll notice equations that have an angle in common. For example, ($\angle A + \angle B = 180°$) and ($\angle B + \angle C = 180°$) both refer to $\angle B$. This allows us to use algebra skills to make some deductions.

What if we subtract $\angle B$ in both of those two equations? Note: to make this easier to read, we'll refer to the angles by their letters, and omit the angle symbol.

$A + B = 180°$

$A + B - B = 180° - B$ Subtract B from both sides of the equation.

$A = 180° - B$

$B + C = 180°$

$B - B + C = 180° - B$ Subtract B from both sides of the equation.

$C = 180° - B$

Therefore, we've proved that A and C are equal both equal the quantity ($180° - B$).

This triangular sail happens to be a version of a 5-12-13 right triangle. If you recognized the Pythagorean triple, you would know that the hypotenuse must be 130 feet long. 130 + 120 + 50 = 300 feet. You also could have used the Pythagorean theorem: $a^2 + b^2 = c^2$, where a and b are the two legs of the triangle and c is the hypotenuse. Choice (B) is the correct answer.

In the following example, you'll encounter a formal proof.

EXAMPLE **7**

In the figure above, the measure of ∠*ABC* is 80°, and ∠*ABC* is bisected by line segment *BD*. What is the measure of ∠*ABD*?

If you were asked to explain your answer to this problem, you'd probably say something like:

"If segment *BD* bisects angle *ABC*, that means it divides the angle exactly in half. Therefore, each of the two halves must equal 40°. The measure of ∠*ABD* is 40°."

If you said that, you'd be correct! However, if you were asked to *write a proof* for this problem, then your explanation should be different. Here's how a proof would look for this example.

You'll often find that there are many possible ways to support a particular proof. That's normal! There's rarely just one "right" answer when it comes to proofs.

Given: ∠*ABC* = 80°
           BD bisects ∠*ABC*

Prove: ∠*ABD* = 40°

Notice the format of this problem. In a proof exercise, you'll often see included information labeled **given**, and a conclusion statement labeled **prove**. Your job is to document the logical steps that lead from the "given" information to the conclusion.

In writing your proof, it's conventional to use two columns, with "Statements" on the left and "Reasons" on the right. You'll usually begin by including some or all of the "given" information in your first statement(s). Also, if no figure has been provided, you should always draw one yourself, and label all of the given information in your figure.

Statements	Reasons
1. ∠*ABC* = 80° *BD* bisects ∠*ABC*	1. Given

Then, you would continue the proof by writing statements that can be derived from the given information. For each statement on the left, you must write your *reason* or justification for that statement on the right. Reasons are always general theorems or postulates—don't use specific names from your figure in the "reasons" column.

Here's how a complete proof might look for this example:

Statements	Reasons
1. $\angle ABC = 80°$ BD bisects $\angle ABC$	1. Given
2. $\angle ABD = \angle ACD$	2. Angle bisector divides an angle into two equal parts
3. $\angle ABD = \angle ABC \div 2$	3. Division produces equal parts
4. $\angle ABD = 80° \div 2$	4. Substitution
5. $\angle ABD = 40°$	5. Division

In a proof, the order of your statements can vary, as long as every statement logically follows from something above it. Additionally, you must never skip steps when writing a formal proof—if a particular fact leads to your next statement, then you must include it in writing, even if it's very obvious. Some instructors are stricter than others when it comes to obvious steps, but it's better to be safe than sorry. Finally, you do not need to include *every* possible fact about the figure in your proof—only those that are necessary and sufficient to reach the conclusion. For example, in this problem, we didn't need to specifically mention that $\angle ACD$ also equals 40°, even though that can be proved as well.

EXAMPLE **8**

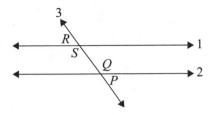

Given that 1 and 2 are parallel lines, explain why ∠P = ∠R.

Often, you'll use known theorems to support your proofs. If you're writing a formal proof, you can refer to theorems by name, or restate the theorems themselves in your Reasons column.

In this example, we know that ∠P + ∠Q = 180°, because they form a linear pair. It's also true that ∠R + ∠S = 180°.

We also know that ∠Q = ∠S, because of the **alternating interior angles theorem**, which states that when two parallel lines are intersected by a transversal, the pairs of alternating interior angles are congruent.

If ∠Q = ∠S, then we can use substitution. Let's substitute ∠Q for ∠S in one of the two linear pairs, so that ∠Q appears in both equations.

∠P + ∠Q = 180°   ∠R + ∠S = 180°

becomes

∠P + ∠Q = 180°   ∠R + ∠Q = 180°

Let's subtract ∠Q from both equations.

∠P = 180° – ∠Q   ∠R = 180° – ∠Q

Therefore, we've proved that ∠P and ∠R are equal—both equal the quantity (180° – ∠Q).

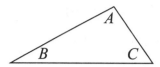

Given: *A*, *B*, and *C* are interior angles of a triangle.

Prove: $\angle A + \angle B + \angle C = 180°$

In this example, we'll explain why the sum of interior angles of a triangle is always 180°. One way to set this up is to draw a line that's parallel to one of the triangle's sides, which lets us use the alternate interior angles postulate.

Let's draw a line through *A* that's parallel to side *BC*. Then, we'll extend side *AB* so that it forms a transversal line (shown below).

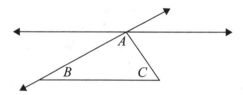

Because we've specified that there are two parallel lines, then we know that the alternate interior angles are congruent. Therefore, we can label the alternate interior angle that's congruent to *B*.

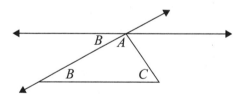

We can use the same process with the other side. Extend *AC* to make a transversal line. Then, we can label the alternate interior angle that's congruent to *C*.

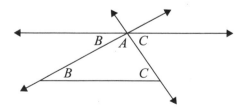

Finally, observe that angles *A*, *B*, and *C* form a straight angle at the top of the figure. Straight angles are always equal to 180°, so these three adjacent angles must have a sum of 180°. Therefore, $\angle A + \angle B + \angle C = 180°$.

EXAMPLE 10

Now that you've seen a few examples of proofs, can you identify the missing step in the proof below?

Given: *A*, *B*, *C*, and *D* are interior angles of a parallelogram.

Prove: $\angle A = \angle C$

Statements	Reasons
1. *ABCD* is a parallelogram	1. Given
2. Segments *AB* and *CD* are congruent	2. Definition of a parallelogram
3. Segments *BC* and *AD* are congruent	3. Definition of a parallelogram
4. Construct segment *BD*	4. Construction
5. Segment *BD* is congruent to segment *BD*	5. Reflexive property of equality
6. _____	6. SSS triangle congruence postulate
7. $\angle A \cong \angle C$	7. Corresponding angles of congruent triangles are congruent.

What is the missing step in the proof above? We gave you a hint in the "reasons" column—we're using the SSS triangle congruence postulate. After constructing segment *BD* according to the proof, which two triangles are formed? The two triangles are *ABD* and *BCD*. (It's fine if you named them a little differently, for example *BDA* for the first triangle, as long as you include the correct 3 vertices). In the previous steps of the proof, we've proven that the corresponding sides of those two triangles are congruent. Therefore, we can use the SSS congruence postulate to say that the two triangles are congruent.

The missing statement is "Triangles *ABD* and *BCD* are congruent."

Which triangle congruence postulate can be used to support statement 7 in the proof below?

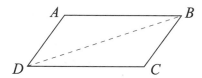

Given: *ABCD* is a parallelogram.

Prove: side *AB* ≅ side *CD*

Statements	Reasons
1. *ABCD* is a parallelogram	1. Given
2. *AB* ∥ *CD*	2. Definition of a parallelogram
3. *BC* ∥ *AD*	3. Definition of a parallelogram
4. ∠*ABD* ≅ ∠*BDC*	4. Alternate interior angles are congruent.
5. ∠*ADB* ≅ ∠*DBC*	5. Alternate interior angles are congruent.
6. Segment *BD* is congruent to segment *BD*	6. Reflexive property of equality
7. Triangles *ABD* and *BCD* are congruent.	7. _____
8. *AB* ≅ *CD*	8. Corresponding angles of congruent triangles are congruent.

What is the support for statement 7? We're claiming that triangles *ABD* and *BCD* are congruent. In the previous statements, we've shown two pairs of corresponding angles are congruent, and the side between them (*BD*) is congruent. The triangle congruence postulate that works here is the ASA (Angle-Side-Angle) congruence postulate.

Lesson 2.3
Constructing Lines and Angles

These days, if we want to create a perfectly precise image with circles, straight lines, or other shapes, we could just use a computer program. However, throughout history, when people needed to be precise, they had to get creative by using instruments. A **compass** is an instrument that aids in drawing perfect circles of almost any size. A **straightedge** aids in drawing perfectly straight-line segments. You'll most likely use a ruler as your straightedge. However, keep in mind that when drawing compass-and-straightedge constructions, you should not be using measuring tools, such as the markings on your ruler. You can find compasses and rulers in the school supply section of any well-stocked department store.

Supplies

For this section, you should have your compass, straightedge, and scratch paper ready. You should be actively practicing the constructions as you read.

Geometry classes today still ask students to use compass and straightedge constructions. This is actually a great way to gain a kinesthetic understanding of concepts such as bisector, perpendicular, and congruence.

CONSTRUCTING LINE SEGMENTS

• •

The first basic exercise is to draw a straight line through two points. This may seem trivial to accomplish; however, some people have trouble getting things lined up exactly right, which is mostly due to the thickness of a typical ruler. Practice getting your line to cross the points precisely. Keep your pencil sharp for these exercises, and angle the pencil lead so that there's no gap between the lead and the edge of the ruler.

Some compasses are cheaply made, and frustrating to use. A good compass will not have wobbly parts, and will hold its radius and pencil lead quite firmly when you use it. If you're having a hard time making circles, the problem might be your compass, and not you!

PERPENDICULAR LINE SEGMENTS

Supplies

If you prefer not to draw directly in this book, access your student tools and download larger, printable versions of the images.

In this exercise, you'll use your compass and straightedge to construct two line segments that are precisely perpendicular to each other. If you've never used a compass before, it may take some practice to get the hang of it. If needed, try making some practice circles on your own first.

Method 1—Start with a Point

It's important to label points where indicated in the instructions, because you're going to connect these points later.

A•

Begin with a point, as in the figure above. For the purposes of this exercise, label the point *A*. Position the compass so that the needle is on point *A*, and make a circle that is perhaps a couple of inches in diameter.

Note that with any of these constructions, it's generally better to use larger circles when possible, as ultra-small circles make it more difficult to find precise intersection points. Strike a balance between reasonable size and clarity in your figures.

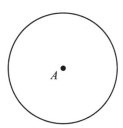

Now, choose any point on the circumference of the circle, and label the point *B*.

Note: You could have also started with two points, *A* and *B*. You would make a circle by placing the compass needle on *A*, and the drawing point on *B*. Either way, at this step you'll have a figure like the one below:

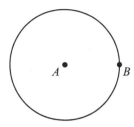

Next, position the compass so that the needle is on point *B*, and the drawing point is on *A*. Try to be precise! With this radius, make a second circle, which will intersect the first circle in exactly two places, points *C* and *D*. Note that this circle has exactly the same radius as circle *A*.

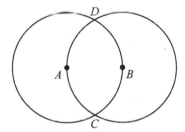

Finally, use your straightedge to draw a line that connects points *A* and *B*. Also draw a line that connects the two intersection points of the circles (points *C* and *D* in the figure above).

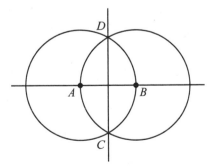

The two lines are precisely perpendicular to each other. One reason this works is that each pair of adjacent points is the same distance apart, as shown in the figure below. Those four points would actually form a **rhombus**, and by definition, a rhombus has diagonals that are perpendicular to each other.

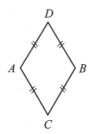

Method 2—Start with a Line Segment

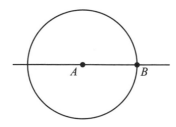

This method is essentially the same as Method 1, except that we begin with a line segment instead of just a point. Begin with a straight-line segment like the one above, with a point labeled A on the line segment. Using your compass, make a circle centered at point A.

This circle intersects the line segment in exactly two places. Choose either one of these intersection points and label it B.

Next, position the compass so that the needle is on point *B*, and the drawing point is on *A*. With this radius, make the second circle, which will intersect the first circle at exactly two places, points *C* and *D*.

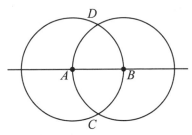

Finally, use your straightedge to draw a line between those two intersection points.

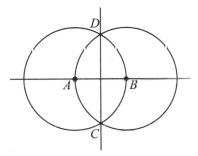

The two lines are precisely perpendicular to each other.

Once you're comfortable with these constructions, you may be able to avoid drawing complete circles with the compass, but instead draw partial arcs at the predicted point(s) of intersection. You would follow the same steps above, but "eyeball" the figure and only draw arcs at the relevant spots on the figure. This is a totally optional adjustment, and may take practice.

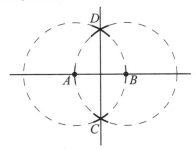

From here on out, we'll show the full circles with dashes and the relevant "arc spots" highlighted in bold.

Try a few more for practice! Follow the steps as above to construct perpendicular lines. Repeat until you're comfortable with the process.

PERPENDICULAR BISECTOR OF A LINE SEGMENT

Next, you'll use your compass to construct a **perpendicular bisector** of a line segment. This is the same as the previous exercise, but you'll use the endpoints of the line segment (A and B in the figure) as your compass points. Use this approach when you need your new perpendicular line to intersect at the exact midpoint of the first line. Begin with a line segment like the one above, with the endpoints labeled A and B. Position the compass so that the needle is on A, and the drawing point is on B, and make a circle.

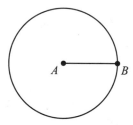

Next, position the compass so that the needle is on point B, and the drawing point is on A. With this radius, make the second circle, which will intersect the first circle at exactly two places, points C and D.

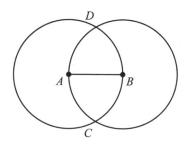

Finally, use your straightedge to draw a line between those two intersection points.

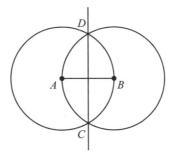

The two lines are precisely perpendicular to each other. Additionally, the new perpendicular line intersects the exact midpoint of the line segment *AB*.

Try a few more for practice! Follow the steps as above to create a perpendicular bisector of a line segment. Repeat until you're comfortable with the process.

PERPENDICULAR LINE THROUGH A POINT

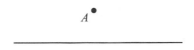

Next, you'll use your compass to construct a line that is perpendicular to an existing segment, and also passes through a given point. This is a little like the perpendicular bisector construction, but in reverse. Use this approach whenever you need a perpendicular line to go through a particular point in the figure.

Begin with a line or segment like the one above, and a point *A* **not** on the line. Position the compass so that the needle is on *A*, and make a circle that intersects the line in two places, points *C* and *B*. We'll call these intersection points *B* and *C*.

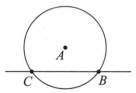

Then, place the compass needle point on *B*, and the drawing point on *A*. Make a circle.

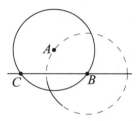

Then, place the compass needle point on *C*, and the drawing point on *A*. Make a circle.

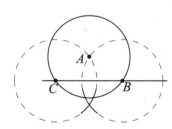

Finally, use your straightedge to draw a line between those two intersection points.

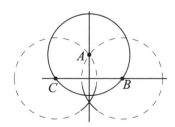

The two lines are precisely perpendicular to each other, and they pass through the point *A*.

Try a few more for practice! Follow the steps as above to create a perpendicular line through a point. Repeat until you're comfortable with the process.

ANGLE BISECTOR

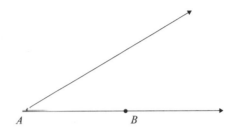

In this exercise, you'll construct a line segment that precisely **bisects** an existing angle. Begin with an angle like the one above (drawn using straightedge), with the vertex labeled *A*, and another point labeled *B* on one of the angle's legs. Position your compass so that the needle point is on *A*, and the drawing point is on point *B*. Turn the compass to make a circle, or just an arc that crosses both of the angle's legs. Each of these two intersection points will be equidistant from the angle's vertex. Label other intersection point *C*.

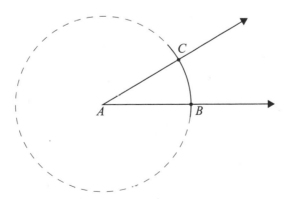

Now, position the compass needle point on *B* and the drawing point on the vertex *A*. With this radius, make a circle, or just an arc that spans the angle's interior.

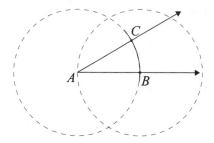

Then, position your compass with the needle point on *C* and the drawing point on the vertex *A*. With this radius, make a circle, or just an arc that spans the angle's interior.

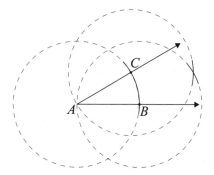

Finally, use your straightedge to draw a line from this new intersection point (labeled *D* in the figure below) to the angle's vertex.

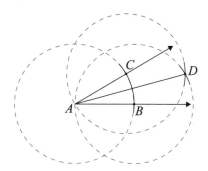

This new line is a bisector of the original angle—it divides the angle precisely in half. One way to prove this is with the SSS triangle congruence theorem. If we connect points to create triangles *ABD* and *ACD*, these triangles would be congruent, because they have three pairs of corresponding congruent sides.

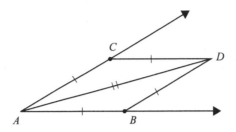

Try a few more for practice! Follow the steps as above to construct an angle bisector. Repeat until you're comfortable with the process.

COPY AN ANGLE

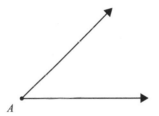

Here, you'll construct an angle that is congruent to another angle. Begin with an angle like the one above, with the vertex labeled *A*. Then, use your straightedge to draw another line segment nearby. This line segment will be one of the legs of the copied angle. Draw a point on the line segment, and label it *A'*.

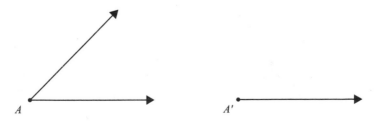

Next, position the compass needle point on the vertex *A*, and make an arc that intersects both of the angle's legs. The arc can be any size. Label the intersection points *B* and *C*.

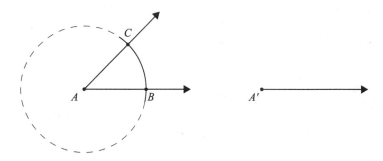

For this step, it's important to carefully hold the compass to the same radius as for the arc above. Position the needle point at *A'*, and make a circle, or just a large arc that intersects the line segment. Label the intersection point *B'*.

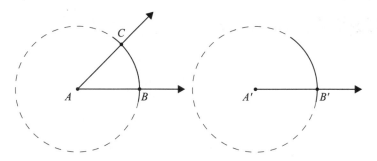

Next, use your compass to "measure" the distance between points *B* and *C*. Position the compass needle point at *B*, and the drawing point at *C*. Carefully holding the compass, position the needle at point *B'*, and draw an arc that intersects the other arc in your copy figure.

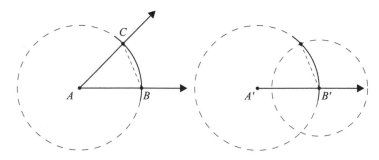

Finally, use your straightedge to draw a line through point *A'* and the intersection of the two arcs. This new, copied angle will be congruent to angle *A*.

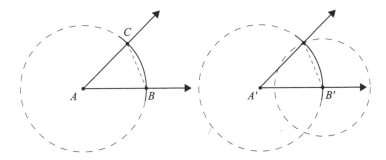

Try a few more for practice! Follow the steps as above to construct an copy of an angle. Repeat until you're comfortable with the process.

PARALLEL LINE THROUGH A POINT

In this exercise, you'll construct a line that is parallel to another line, and also goes through a specified point. This method is based on our knowledge of parallel lines and transversals—that when two parallel lines are intersected by a transversal line, the alternate interior angles are congruent. Here, we'll copy an angle, using the strategy from the previous exercise, in order to create a parallel line.

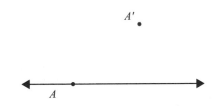

Begin with a line like the one above, with a point somewhere on the line (labeled *A*) and a point not on the line (labeled *A'*). Use your straightedge to draw a line segment that intersects *A* and *A'*. Make sure to extend the line a little beyond point *A'* so that you have space to copy the angle there later.

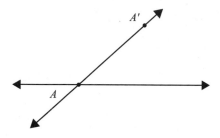

Next, using the strategy from the previous exercise, copy the angle formed by these two lines. Position the compass needle point on the vertex *A*, and make an arc that intersects the two line segments. Label the intersection points *B* and *C*.

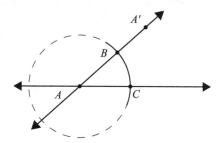

Carefully holding the compass to the same radius, position the needle at point *A'*. Draw an arc, and label the intersection point *B'*.

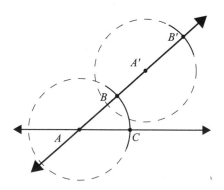

Next, use your compass to "measure" the distance between points *B* and *C*. Then, carefully holding the compass to the same radius, position the needle at point *B'*, and draw an arc that intersects the other arc in your copy figure.

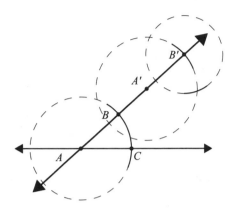

Finally, use your straightedge to draw a line through point *A'* and the intersection of the two arcs. This new, copied angle will be congruent to angle *A*.

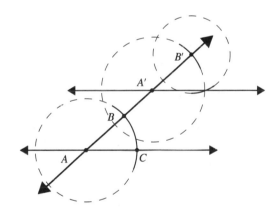

Try a few more for practice! Follow the steps as above to construct a parallel line through a point. Repeat until you're comfortable with the process.

Lesson 2.4
Constructing Polygons

EQUILATERAL TRIANGLE

Method 1—Start with Two Points

Supplies

For this section, you should have your compass and straightedge ready.

In this exercise, you'll use your compass and straightedge to construct an **equilateral triangle**. The beginning steps are the same as the methods for creating **perpendicular lines**, seen earlier in this chapter. Therefore, we'll keep the instructions brief. Feel free to refer to the earlier section on constructing perpendicular lines for more details.

$A \bullet$ $\bullet B$

Constructing an equilateral triangle is very much like constructing perpendicular lines! (See earlier in Lesson 2.3.)

Position the compass so that the needle is on point A, and the drawing point is on B. Make an arc above the two points, in the spot where the third vertex of the triangle should go (eyeball it). You can also make a full circle if you prefer.

Next, position the compass so that the needle is on point *B*, and the drawing point is on *A*. With this radius, make a second arc to intersect the first one. Label this point *C*.

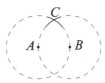

Finally, connect points *A*, *B*, and *C* to form a triangle. Use your straightedge.

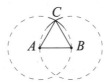

This triangle is equilateral. The reason this works is that the compass radius always matched the distance between *A* and *B*, so we know that all three sides are equal.

Try a few more for practice! Follow the steps as above to create an equilateral triangle. Repeat until you're comfortable with the process.

Method 2—Inscribed in a Circle

For this method, begin with a circle, with a point marked for its center. Label the center point *A*. Also choose a point on the circle's circumference, and label it *B*.

Position the compass so that the needle is on point *B*, and the drawing point is on *A*. With this radius, make two arcs that intersect the original circle. For the purposes of this exercise, label these intersection points *C* and *G*.

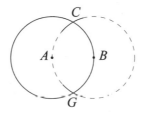

Next, position the compass so that the needle is on point *C*, and the drawing point is on *A*. With this radius, make an arc that intersects the original circle. Label this intersection point *D*.

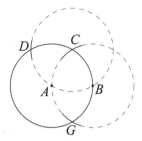

Next, position the compass so that the needle is on point *D*, and the drawing point is on *A*. With this radius, make an arc that intersects the original circle. Label this intersection point *E*.

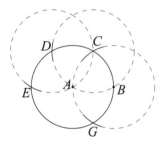

Next, position the compass so that the needle is on point *E*, and the drawing point is on *A*. With this radius, make an arc that intersects the original circle. Label this intersection point *F*.

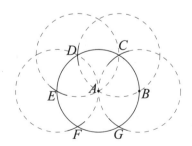

The final step is to connect points *B*, *D*, and *F* to make a triangle. (Alternatively, you could use points *C*, *E*, and *G* instead.)

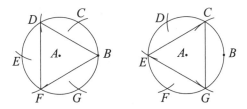

This triangle is equilateral, and it is **inscribed** in the circle. The reason this works is that we used the same compass radius for each arc we made—in other words, the intersection points were equally spaced from each other.

REGULAR HEXAGON

This method uses all the same steps as those used in the previous exercise (equilateral triangle inscribed in a circle). The difference is that at the end, you'll connect all six intersection points instead of just three.

Follow the steps in the previous exercise (equilateral triangle inscribed in a circle), proceeding until you've drawn all six arcs on the circle. See the figure below.

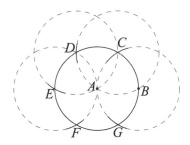

The final step is to connect these six intersection points to form a hexagon, as in the figure below. Use your straightedge.

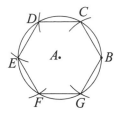

This hexagon is equilateral, and it is **inscribed** in the circle. The reason this works is that we used the same compass radius for each arc we made—in other words, the intersection points were equally spaced from each other.

SQUARE

Method 1—Start with Perpendicular Lines

In this exercise, you'll use your compass and straightedge to construct a **square**. For this method, start by following the steps for constructing **perpendicular lines**, seen earlier in this chapter. You can use either of the methods recommended. It might be a good idea to erase the circles as best you can, leaving only the perpendicular lines.

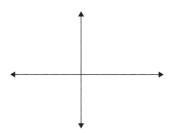

Once you have perpendicular lines, proceed to the next step. Position your compass needle at the intersection point of the two lines, and make a circle. You can also just make arcs at the points where the circle crosses the two perpendicular lines.

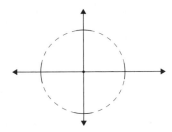

The circle crosses the two perpendicular lines, creating four intersection points. The next step is to connect those four intersection points to form a square, as in the figure below.

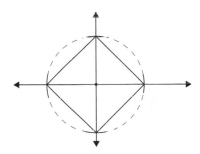

The quadrilateral drawn is a perfect square. One reason this works is that we know the two diagonals are perpendicular (we constructed them that way), and the four vertices are equidistant from the center.

Try a few more for practice! Follow the steps as above to create a square. Repeat until you're comfortable with the process.

Method 2—Inscribed in a Circle

For this method, begin with a circle, with a point marked for its center.

Position your straightedge carefully so that its line passes through the circle's center. Draw a straight line, which will be the diameter of the circle, or just mark the points where the straightedge intersects the circle. Label these points A and B.

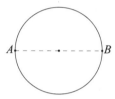

The next steps will be to construct a perpendicular bisector of this circle's diameter. Position the compass so that the needle is on point A, and the drawing point is on B. Make a circle, which will be larger than the given circle. Or, just make two arcs, one above and one below the diameter, spanning the spots where the perpendicular bisector will go (eyeball it).

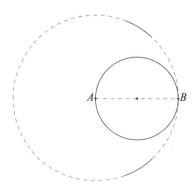

Next, position the compass so that the needle is on point *B*, and the drawing point is on *A*. Make a circle, or just two arcs, which intersect the previous arcs.

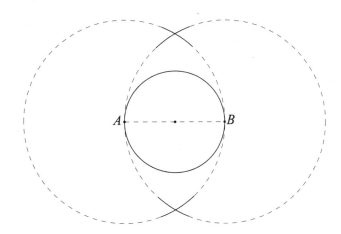

Position your straightedge to connect the two intersections of these larger arcs (it should also pass through the circle's center). Draw a straight line, or just mark the points where the straightedge intersects the original circle. Label these points *C* and *D*.

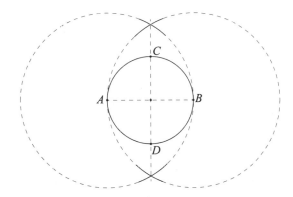

Finally, use your straightedge to connect points *A*, *B*, *C*, and *D*, forming a square.

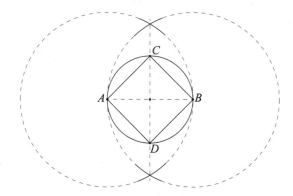

This square is **inscribed** in the circle.

Try a few more for practice! Follow the steps as above to create a square. Repeat until you're comfortable with the process.

ANSWERS TO CHAPTER 2 EXERCISES

SSS Postulate Exercise (Page 53)

1. No.
   These triangles have different lengths for all three sides; therefore, they are not congruent.

2. No.
   These triangles have only one side length in common; therefore, they are not congruent.

3. No.
   These triangles have two side lengths in common. However, the third side is not the same for both triangles. Therefore, they are not congruent.

4. Yes.
   These triangles have the same lengths for all three sides; therefore, they are congruent.

5. Yes.
   These triangles have the same lengths for all three sides; therefore, they are congruent.

6. No.
   These triangles have two side lengths in common. However, one side length is unknown; therefore, they are not necessarily congruent.

AAS Postulate Exercise (Page 57)

1. No.
   These triangles have different lengths for the given sides; therefore, they are not congruent.

2. Yes.
   Use the ASA postulate—there is an angle, side, and angle (in that order) that are congruent in the two triangles. Therefore, they are congruent.

3. No.
   These triangles have two angles in common. However, the given side is in a different place relative to the angles—in the first triangle, the given side is opposite the 30° angle, and in the second triangle, it's opposite the 50° angle. Therefore, they are not congruent.

4. Yes.
Use the AAS postulate—there is an angle, angle, and side (in that order) that are congruent in the two triangles. Therefore, they are congruent.

5. No.
These triangles have one side and one angle in common. However, they do not follow either ASA or AAS congruence. Therefore, they are not necessarily congruent.

6. Yes.
Use the AAS postulate—there is an angle, angle, and side (in that order) that are congruent in the two triangles. Therefore, they are congruent.

CHAPTER 2 PRACTICE QUESTIONS

Directions: Complete the following problems as specified by each question. For extra practice after answering each question, try using an alternative method to solve the problem or check your work. Larger, printable versions of images are available in your online student tools.

1. Which of the following could represent the side lengths of a right triangle? (Select all that apply.)

   ☐ 30, 60, 90

   ☐ 15, 36, 39

   ☐ 30, 40, 50

   ☐ 16, 30, 34

   ☐ 15, 21, 27

   ☐ 14, 48, 51

2. Using constructions, and not direct measurement, determine whether the triangle below is equilateral.

   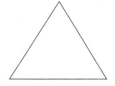

3. Using constructions, and not direct measurement, determine whether $AC \cong BC$.

   C

   A•————————————•B

4.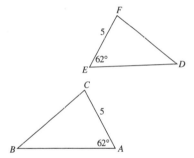

   Which of the following facts alone, if true, would be sufficient to prove that triangles *ABC* and *DEF* are congruent? (Select all that apply.)

   ☐ $AB \cong ED$

   ☐ $AC \cong FD$

   ☐ $CA \cong ED$

   ☐ $\angle ACB \cong \angle DFE$

   ☐ $\angle CBA \cong \angle FDE$

5. Given: $\overline{DE} \perp \overline{EF}$
$\overline{XY} \perp \overline{YZ}$

Prove: $m\angle DEF = m\angle XYZ$

Complete the following proof:

Statements	Reasons
1. $\overline{DE} \perp \overline{EF}$	1. Given
2. $\angle DEF$ is a right angle	2.
3.	3. A right angle has a measure of 90°.
4.	4. Given
5.	5. Definition of perpendicular
6. $m\angle XYZ = 90$	6.
7.	7. Transitive property of equality (steps 3 and 6)

6.

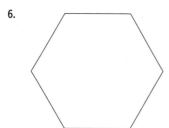

The figure above is a regular hexagon with side length 6. Which of the following triangles could fit inside the hexagon? (The triangle may be rotated, and it does not necessarily have to be inscribed).

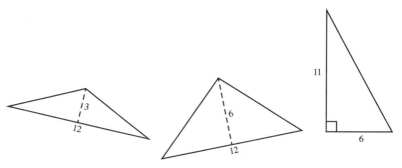

7. A triangle has sides with lengths 14 and 6. Find:

    a. the range of possible values for the third side

    b. the range of possible values for the perimeter

8.

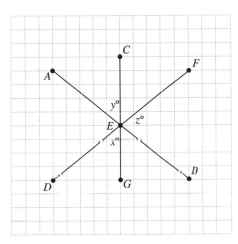

In the figure provided, $\overline{EF}$ bisects $\angle CEB$, DE bisects $\angle AEG$, and AB, CG, and DF all intersect at E.

    a. If $x = 50°$, then find the value of all other angles.

    b. Does $2(x + y + z) = 360°$?

    c. Is it always true that $x = z$?

    d. Is it always true that $y = x$?

SOLUTIONS TO CHAPTER 2 PRACTICE QUESTIONS

1. **(B), (C), and (D) are right triangles.**

   For each example, use Pythagorean theorem ($a^2 + b^2 = c^2$) to determine whether the numbers could represent the side lengths of right triangles. Since there are no unknowns (variables) in these equations, you're just trying to determine whether the equations are true.

   Reminder: The longest side is *always* the hypotenuse, or c in the equation.

   ☒ $30^2 + 60^2 = 90^2$ Pythagorean theorem with $a = 30$, $b = 60$, and $c = 90$
      $900 + 3600 = 8100$ Simplify
      $4500 \neq 8100$ ✗ Not a true equation. This is not a right triangle.

   ☑ $15^2 + 36^2 = 39^2$ Pythagorean theorem with $a = 15$, $b = 36$, and $c = 39$
      $225 + 1296 = 1521$ Simplify
      $1521 = 1521$ ✔ True equation. This is a right triangle.

   Note: 15:36:39 is the third multiple of 5:12:13, a common Pythagorean triple.

   ☑ $30^2 + 40^2 = 50^2$ Pythagorean theorem with $a = 30$, $b = 40$, and $c = 50$
      $900 + 1600 = 2500$ Simplify
      $2500 = 2500$ ✔ True equation. This is a right triangle.

   Note: 30:40:50 is the tenth multiple of 3:4:5, a common Pythagorean triple.

   ☑ $16^2 + 30^2 = 34^2$ Pythagorean theorem with $a = 16$, $b = 30$, and $c = 34$
      $256 + 900 = 1156$ Simplify
      $1156 = 1156$ ✔ True equation. This is a right triangle.

   Note: 16:30:34 is the second multiple of 8:15:17, a common Pythagorean triple.

   ☒ $15^2 + 21^2 = 27^2$ Pythagorean theorem with $a = 15$, $b = 21$, and $c = 27$
      $225 + 441 = 729$ Simplify
      $666 \neq 729$ ✗ Not a true equation. This is not a right triangle.

   ☒ $14^2 + 48^2 = 51^2$ Pythagorean theorem with $a = 14$, $b = 48$, and $c = 51$
      $196 + 2304 = 2601$ Simplify
      $2500 \neq 2601$ ✗ Not a true equation. This is not a right triangle.

2. The triangle is equilateral.

In Lesson 2.4, the way we constructed an equilateral triangle was by using circles centered at two points, similar to constructing a perpendicular bisector. The third vertex was the intersection of these two circles.

You can use the same idea to determine whether a triangle is equilateral. Make two circles, each with radius *AB*, and centered at *A* and *B,* respectively. (Points are labeled here for the purposes of the explanation). Since *C* lies on the intersection point of the two circles, the triangle is equilateral.

Another way to think about this is that each of the three sides is equal to the radius of the circles. *AB* is the radius of both circle *A* and circle *B*, so we know that both circles are congruent. *AC* is the radius of circle *A*, and *BC* is the radius of circle *B*.

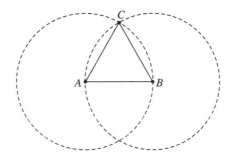

3. It is true that *AC* ≅ *BC*.

One efficient way to solve this problem is to construct a perpendicular bisector through segment *AB*. Since a bisector divides a line segment precisely in half, this will tell you whether *AC* ∼ *BC*.

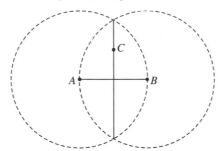

Construct the perpendicular bisector as shown above. Make two circles, each with radius *AB*, and centered at *A* and *B,* respectively. The intersection of these two circles is the perpendicular bisector. You should find that it passes through point *C*. Since *AC* lies on the perpendicular bisector, you know that *AC* ≅ *BC*.

Another explanation for this figure is in the form of right triangles. If you drew the segments connecting *AC* and *BC*, you would have two congruent right triangles. You know that they are congruent, because segment *AB* is bisected, and the other segment through point *C* is shared between both triangles, so that segment is congruent to itself. That means that the hypotenuses of these triangles must be congruent as well.

4. ☑ **$AB \cong ED$**

    ☒ $AC \cong FD$

    ☒ $CA \cong ED$

    ☑ **$\angle ACB \cong \angle DFE$**

    ☑ **$\angle CBA \cong \angle FDE$**

$AB \cong ED$ would prove congruence through SAS.

$AC \cong FD$ would not be sufficient to prove congruence, because SSA is not a congruence property. Actually, if you *knew* that both of these triangles are acute (as they appear to be), then SSA can work (since it rules out one of the two possible triangles that can be made from that information). These triangles appear to be acute, but you should not assume that figures are drawn to scale unless they are specifically described as such. So based on the given information alone, you should conclude that the triangles can't be proved congruent.

$CA \cong ED$ would not be sufficient to prove congruence, because the segments do not correspond. (CA is opposite the given angle, while ED is adjacent to the given angle).

$\angle ACB \cong \angle DFE$ would prove congruence through ASA.

$\angle CBA \cong \angle FDE$ would prove congruence through AAS.

5. Your proof should look like the one below. It's fine if you have some slight differences that are equivalent in meaning.

Statements	Reasons
1. $\overline{DE} \perp \overline{EF}$	1. Given
2. $\angle DEF$ is a right angle	2. **Definition of perpendicular**
3. **m$\angle DEF$ = 90°**	3. A right angle has a measure of 90°.
4. $\overline{XY} \perp \overline{YZ}$	4. Given
5. **$\angle XYZ$ is a right angle**	5. Definition of perpendicular
6. m$\angle XYZ$ = 90°	6. **A right angle has a measure of 90°.**
7. **m$\angle DEF$ = m$\angle XYZ$**	7. Transitive property of equality (steps 3 and 6)

For Statement 2, we're saying that $\angle DEF$ is a right angle. How do you know that it is a right angle? Because you know that the segments are described as perpendicular (Statement 1), and you know that means that the segments intersect at a right angle. Thus, the reason is the **definition of perpendicular**. You may have also cited the actual definition, i.e. "Perpendicular segments intersect at right angles."

For Statement 3, choose a statement that is related to the previous statements and/or given information, and is proved with the *reason* "a right angle has a measure of 90°." The relevant statement here is that **m∠DEF = 90°**. This may have seemed like an overly obvious statement to make; but remember, for strict formal proofs, it's necessary to include reasoning for *every* step in your logic.

For Statement 4, we know that we need a <u>"given"</u> statement, since that is the *reason* that is cited. The other "given" statement is that **XY ⊥ YZ**. Note that this also flows nicely with the logic of the proof, since the next statements are about m∠XYZ.

Statements 5 and 6 use the same logic as statements 2 and 3. You can prove that ∠XYZ is a *right angle* based on the "definition of perpendicular." Then, prove that m∠XYZ = 90° based on the fact that **a right angle has a measure of 90°**. For Statement 6, you could have perhaps written "definition of right angle," rather than citing the definition itself.

The final statement should always be the *PROVE* statement. In this case, you were trying to prove that **m∠DEF = m∠XYZ**. Also, you should verify that this statement is proved with the transitive property of equality. The transitive property states that, if you know that different things have the same value, then you know those things are equal to each other. This is a relevant property to use for the final statement, since you have proved that both ∠DEF and ∠XYZ are equal to 90°.

6.

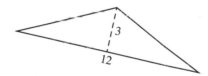

Of the triangles listed, only the first one would fit inside the hexagon.

Consider what you know about the hexagon. The longest diagonal must have a length of 12. (You can think of a hexagon as six smaller equilateral triangles that fit together—in this case, the smaller triangles would have a side length of 6, and two of them side-by-side would have a length of 12.)

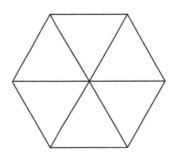

That means that a triangle with a base of 12 can fit in the width of the hexagon. But, how tall can that triangle be? Each smaller equilateral triangle would have a base of 6, and a height of $3\sqrt{3}$. You can use the Pythagorean theorem to find this ($a^2 + b^2 = c^2$, $3^2 + b^2 = 6^2$, solve for b.) The value of $3\sqrt{3}$ is approximately 5.2. Therefore, the first triangle would fit inside the hexagon, but the second would not.

Similarly, if you have a right triangle with a base of 6 (as in the third triangle shown in the question), that would fit along the base of the hexagon, but how tall can it be? In the previous steps, you found that half the altitude of the hexagon is equal to $3\sqrt{3}$. Double that to find the altitude: $6\sqrt{3}$, or approximately 10.4. This is smaller than 11; therefore, the third triangle would not fit in the hexagon, either.

7. **a.** **The third side can have any length between 8 and 20.**
   b. **The perimeter can have any length between 28 and 40.**

Recall the Third Side Rule: *In any triangle, the length of one side must be greater than the difference and less than the sum of the other two sides.*

To find the range for the third side, find the sum and the difference of the two given values, 14 and 6. The sum is $14 + 6 = 20$. The difference is $14 - 6 = 8$. Therefore, the range of values for the third side is $8 < x < 20$.

To find the range for the perimeter, use the values you found for the third side. The third side must be greater than 8, so the perimeter must be greater than 28 ($= 8 + 14 + 6$). The third side must be less than 20, so the perimeter must be less than 40 ($= 20 + 14 + 6$).

8. **a.** Through vertical angle congruency, we know that $\angle AEC \cong \angle GEB$, $\angle CEF \cong \angle DEG$, and $\angle FEB \cong \angle AED$. Next, if $x = 50$, then $\angle DEG = \angle CEF = 50°$. Additionally, $\overline{EF}$ bisects $\angle CEB$ (given), so we know that $\angle CEF \cong \angle FEB$. Therefore, $\angle CEF = \angle FEB = \angle AED = 50°$. Finally, you can prove that $\angle AEC = \angle GEB = 80°$, because $80° + 50° + 50° = 180°$ (use adjacent angles forming a straight line).

   b. Since $\angle AEC \cong \angle GEB$, $x + y + z = 180$, and $\dfrac{360}{2} = 180$; then, through the transitive property, $x + y + z = \dfrac{360}{2}$, and $2(x + y + z) = 360$. Therefore, yes, $2(x + y + z) = 360°$.

   c. Through vertical angle congruency, $\angle DEG \cong \angle CEF$. Since $\overline{EF}$ bisects $\angle CEB$, it is always true that $x = z$.

   d. $\angle AEC \cong \angle GEB$ through vertical angle congruency and $\angle AED \cong \angle DEG \cong \angle CEF \cong \angle BEF$ for bisector properties and vertical angle congruency. Since there is no equation of angles for y and x, they are not ever equal to one another. Therefore, $y \neq x$.

REFLECT

Congratulations on completing Chapter 2! Here's what we just covered. Rate your confidence in your ability to:

- Identify congruent triangles using the SSS, AAS, ASA, SAS postulates

    ① ② ③ ④ ⑤

- Use the Pythagorean theorem or the Third Side Rule to solve for an unknown side in a right triangle

    ① ② ③ ④ ⑤

- Understand how to write a formal or informal proof

    ① ② ③ ④ ⑤

- Construct figures with a compass and straightedge (parallel and perpendicular lines, angle bisector, angle copy, equilateral triangle, square, and regular hexagon)

    ① ② ③ ④ ⑤

If you rated any of these topics lower than you'd like, consider reviewing the corresponding lesson before moving on, especially if you found yourself unable to correctly answer one of the related end-of-chapter questions.

 Access your online student tools for a handy, printable list of Key Points for this chapter. These can be helpful for retaining what you've learned as you continue to explore these topics.

Chapter 3
Similarity

3

GOALS **By the end of this chapter,
you will be able to:**

- Understand and perform dilations
  of figures, including figures in the
  coordinate plane

- Identify similar triangles using
  the SSS, AAA, and SAS similarity
  postulates

Lesson 3.1
Dilations

In this lesson, we will review another type of image transformation—**dilation**. A dilated image is the same shape as the pre-image, but is a different size. In other words, dilation stretches or shrinks the original figure. Additionally, the image and pre-image are **similar** but not **congruent**.

In order to perform a dilation of a figure, we need to know the **scale factor** and the **center of dilation**. The scale factor is the ratio of the corresponding segment lengths of the two figures, and the center of dilation is the point of reference used to orient the figure. Additionally, if you pick any point in the dilated image, you'll be able to draw a straight line through that point, its corresponding pre-image point, and the center of dilation.

> Each point in the dilated image is **collinear** with its corresponding pre-image point and the center of dilation.

EXAMPLE 1

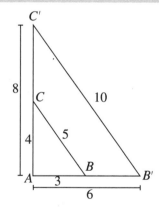

In the figure above, triangle *ABC* is dilated with a scale factor of 2 and the center of dilation is at vertex *A*.

Since the scale factor is 2, that means that the ratio of corresponding segment lengths is 1:2. In other words, the larger triangle has lengths two times greater than the smaller triangle. Therefore, the larger triangle has side lengths of 6, 8, and 10—corresponding with the smaller side lengths of 3, 4, and 5, respectively.

The center of dilation is how we position the vertices of the dilated figure. For each point in the image, the center of dilation is collinear with the image point as well as its corresponding pre-image point. In Example 1, the center of dilation was given as vertex A. We have two sets of collinear vertices: points A, B, and B′ are collinear, and points A, C, and C′ are also collinear.

If a figure has more vertices, we'll most likely use the ratio of some of the **diagonals** for reference, rather than using only the side lengths.

EXAMPLE 2

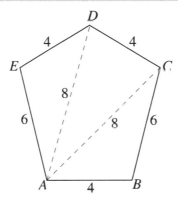

Supplies

Access your student tools to download larger, printable versions of the images in this section.

Construct a dilated version of pentagon *ABCDE*, with scale factor of 1.5 and a center of dilation at vertex *A*.

If the scale factor is 1.5, then each segment in the dilated figure will be 1.5 times the length of the corresponding segment in the pre-image. Therefore, the segments will have the following lengths:

Pre-Image (length)	Image (length)
AB (4)	AB′ (6)
BC (6)	B′C′ (9)
CD (4)	C′D′ (6)
DE (4)	D′E′ (6)
EA (6)	E′A (9)
AC (8)	AC′ (12)
AD (8)	AD′ (12)

We're halfway there! But, how do we position the vertices in the dilated image? Recall that each point in the dilated image must be collinear with its corresponding pre-image point, as well as the center of dilation (in this case, vertex *A*). So, next, we'll just need to extend each of the sides and diagonals that connect to vertex *A*.

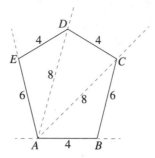

You would then measure each segment so that it matches the lengths shown in the table above. Since this is an example exercise, a scale version of the image is shown here:

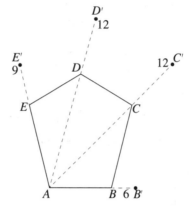

Finally, connect the new vertices to form a pentagon.

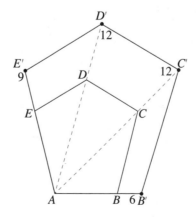

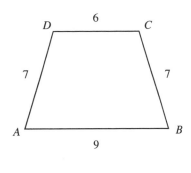

Construct a dilated version of quadrilateral *ABCD*, with scale factor 2 and center of dilation *E*.

This example is different, because the center of dilation is not one of the vertices. Thus, the given side lengths don't help us all that much, because we need to position the dilated figure with respect to point *E*. Nevertheless, the process is quite similar to the previous exercise, with one important change: You're going to measure the **distance** from each vertex to point *E*, and apply the scale ratio to those measurements.

First, draw a line from point *E* through each vertex. Then, measure the distance formed by each of these pairs of points (*FA*, *EB*, *EC*, and *ED*). Since this is an example exercise, we'll provide the "given" distances as follows:

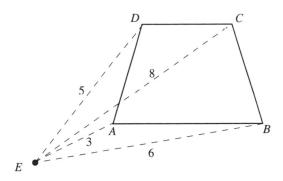

Next, extend these segments (*EA*, *EB*, *EC*, and *ED*) to 2x their length. Remember to start each measurement from point *E*. Label the new endpoints *A'*, *B'*, *C'*, and *D'*.

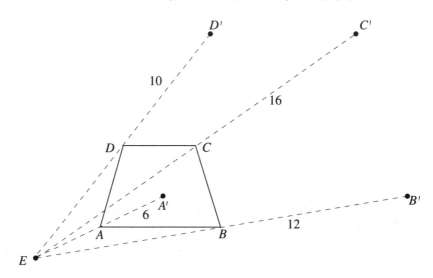

Finally, connect the four new points to form a quadrilateral.

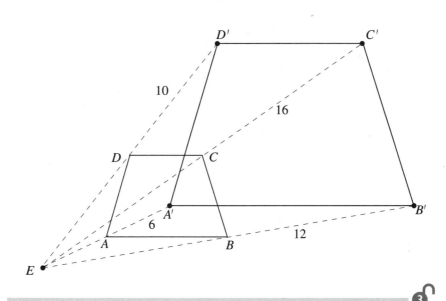

EXAMPLE **4**

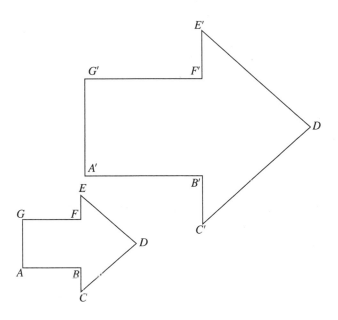

Figure *ABCDEFG* above is dilated with a scale factor of 2. Find the center of dilation.

In this example, you're asked to find the center of dilation. This process is actually quite simple—it just involves drawing some straight lines, and you won't need to measure.

First, choose a pair of corresponding points, for instance, *A* and *A'*. Draw a long, straight line through these points. (Hint: The center of dilation is sort of to the left of point *A*, so extend the line plenty in that direction.)

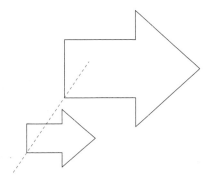

Do the same thing with a different pair of corresponding points, for instance, *B* and *B′*. Draw a long, straight line through these points.

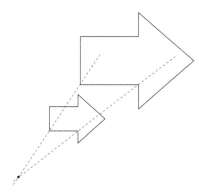

Where these two lines intersect is the center of dilation.

You only need to draw two lines, but it doesn't hurt to add the remaining lines to check your work.

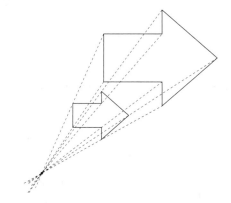

DILATIONS IN THE COORDINATE PLANE

If the center of dilation is the **origin** (point 0, 0), you'll apply the scale factor multiple to each reference point in the figure. For example, if the scale factor is 4, then point (2, 3) would be dilated to (8, 12).

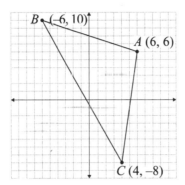

Construct a dilated version of triangle *ABC* with a scale factor of $\frac{1}{2}$ and the center of dilation at (0, 0).

When the center of dilation is at the origin, all you need to do is multiply each coordinate by the scale factor $\left(\frac{1}{2}\right)$. Complete this process for each of the three vertices, and find the image coordinates as follows:

Pre-Image	Image
A (6, 6)	*A'* (3, 3)
B (-6, 10)	*B'* (-3, 5)
C (4, -8)	*C'* (2, -4)

Finally, plot the three points and then connect the three vertices to form a triangle.

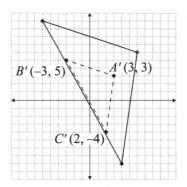

EXAMPLE 6

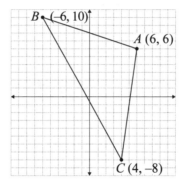

Construct a dilated version of triangle *ABC* with a scale factor of $\frac{1}{2}$ and the center of dilation at (–2, –4).

If the center of dilation is *not* at the origin, then you'll use the same process that we reviewed earlier in this chapter—applying the scale factor multiple to the *distance from the center of dilation*. This is a little easier than it might sound!

Method 1—Counting

Let's begin with vertex A (6, 6). We can apply the same process as we did with non-coordinate dilation, extending lines from the center of dilation and measuring the distance.

Draw a line from the center of dilation and through coordinate A. Of course, the distance wasn't given here, so we'll have to find it. The good news is that we won't need to use Distance Formula or anything overly complicated—we'll just focus on the horizontal and vertical difference between the points. By counting, we can see that coordinate A is **8** horizontal units and **10** vertical units from the center of dilation.

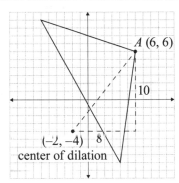

The scale factor for this exercise is $\frac{1}{2}$. That means that the dilated image of A (labeled A' in the figure below) should be **4** horizontal units and **5** vertical units (in the same direction) from the center of dilation. In other words, the distance between A' and (−2, −4) is $\frac{1}{2}$ the distance between A and (−2, −4). The coordinate A' is located at (2, 1).

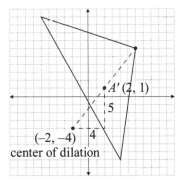

Repeat this process for coordinates *B* and *C*.

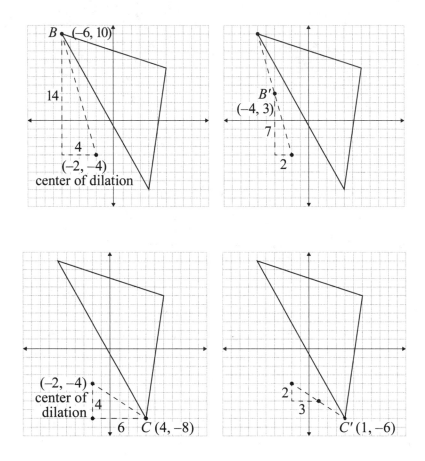

Finally, connect the vertices to form a triangle.

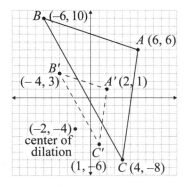

Method 2—Arithmetic

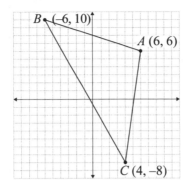

Construct a dilated version of triangle *ABC* with a scale factor of $\frac{1}{2}$ and the center of dilation at (–2, –4).

We can also complete this process arithmetically, instead of drawing and counting.

Let's begin with coordinate *A*. We're going to focus on the distance between *A* and the center of dilation (–2, –4).

Subtract the coordinates of (–2, –4) from the coordinates of (6, 6). You should always subtract in that order—your reference point minus the center of dilation—so that you have the correct signs.

6 – (–2) = 8	Subtract the *x*-coordinates.
6 – (–4) = 10	Subtract the *y*-coordinates.

Multiply each of those two numbers by the scale factor $\left(\frac{1}{2}\right)$.

$$8 \times \frac{1}{2} = 4$$

$$10 \times \frac{1}{2} = 5$$

Add each of those two numbers, respectively, to the *x*- and *y*-coordinates of the center of dilation.

$$(-2) + 4 = 2$$

$$(-4) + 5 = 1$$

These are the coordinates for the dilated version of point *A*.

Point *A'* is at (2, 1).

Apply the same process to the other two vertices.

A (6, 6)	*B (–6, 10)*	*C (4, –8)*	**Original Coordinate**
			Note: The center of dilation is (–2, –4). The scale factor is $\frac{1}{2}$.
$6 - (-2) = 8$ $6 - (-4) = 10$	$(-6) - (-2) = -4$ $10 - (-4) = 14$	$4 - (-2) = 6$ $-8 - (-4) = -4$	Subtract the coordinates (those of your point minus those of the center of dilation, in that order).
$8 \times \dfrac{1}{2} = 4$ $10 \times \dfrac{1}{2} = 5$	$-4 \times \dfrac{1}{2} = -2$ $14 \times \dfrac{1}{2} = 7$	$6 \times \dfrac{1}{2} = 3$ $-4 \times \dfrac{1}{2} = -2$	Multiply each of those two numbers by the scale factor.
$(-2) + 4 = 2$ $(-4) + 5 = 1$	$(-2) + (-2) = -4$ $(-4) + 7 = 3$	$(-2) + 3 = 1$ $(-4) + (-2) = -6$	Add each of those two numbers, respectively, to the *x*- and *y*-coordinates of the center of dilation.
(2, 1)	(–4, 3)	(1, –6)	Those are the dilated versions of the original coordinates.

Finally, plot the vertices and complete the triangle.

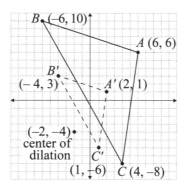

EXAMPLE **7**

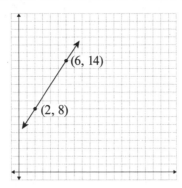

Construct a dilated version of line 1 with a scale factor of 1.5 and the center of dilation at (0, 0).

Method 1—Counting

To form a line, you just need two points. In general, you can choose any two points from the line to work with. In this example, two coordinates are provided, so that makes things easier.

To perform the dilation on these two points, follow the same steps as those in the previous exercise. For point (2, 8), the x-coordinate is 2 units away from the center of dilation, and the y-coordinate is 8 units away from the center of dilation.

For coordinate (6, 14), the x-coordinate is 6 units away from the center of dilation, and the y-coordinate is 14 units away from the center of dilation.

If two points are not provided in the figure, then perhaps an equation for the line will be provided. In that case, plug in two different values for x, and solve for the corresponding y-coordinates.

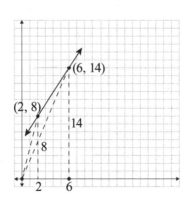

The scale factor for this dilation is 1.5. Therefore, the dilated version of (2, 8) should be 3 horizontal units and 12 vertical units from the center of dilation.

The dilated version of (6, 14) should be 9 horizontal units and 21 vertical units from the center of dilation.

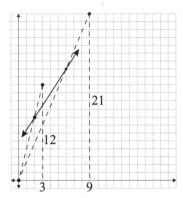

This places the new coordinates at (3, 12) and (9, 21), respectively.

Finally, connect these new coordinates to form a line.

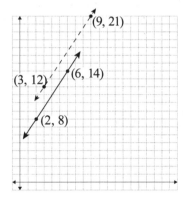

You may notice that these lines appear to be parallel to each other. In fact, this will always be true when performing a dilation of a line!

Dilation of Lines

A dilated version of a line will always be parallel to the original line.

Exception: If the center of dilation *is a point on the line*, then the dilation will leave the line unchanged.

Method 2—Arithmetic

Here is the arithmetic solution for Example 7.

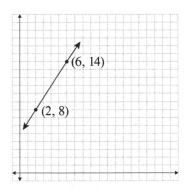

Construct a dilated version of line 1 with a scale factor of 1.5 and the center of dilation at (0, 0).

A (2, 8)	B (6, 14)	Original Coordinate
		Note: The center of dilation is (0, 0).
		The scale factor is 1.5.
2 – 0 = 2 8 – 0 = 8	6 – 0 = 6 14 – 0 = 14	Subtract the coordinates (those of your point minus those of the center of dilation, in that order). (Notice that if the center of dilation is (0, 0), then this step is basically unnecessary.)
2 × 1.5 = 3 8 × 1.5 = 12	6 × 1.5 = 9 14 × 1.5 = 21	Multiply each of those two numbers by the scale factor.
3 + 0 = 3 12 + 0 = 12	9 + 0 = 9 21 + 0 = 21	Add each of those two numbers, respectively, to the x- and y-coordinates of the center of dilation. (Notice that if the center of dilation is (0, 0), then this step is basically unnecessary.)
(3, 12)	(9, 21)	Those are the dilated versions of the original coordinates.

Finally, plot the new coordinates and connect them to form a line.

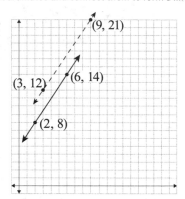

EXAMPLE 8

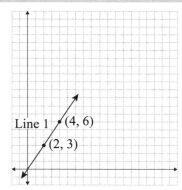

Construct a dilated version of line 1 with a scale factor of 3 and the center of dilation at (0, 0).

Method 1—Counting

You may notice that the center of dilation (0, 0) is a point that lies on our original line. As mentioned previously, this will mean that the dilation leaves the line unchanged! Let's complete this dilation to see it in action.

To perform the dilation on these two points, follow the same steps as in the previous exercise. For point (2, 3), the x-coordinate is 2 units away from the center of dilation, and the y-coordinate is 3 units away from the center of dilation.

For coordinate (4, 6), the x-coordinate is 4 units away from the center of dilation, and the y-coordinate is 6 units away from the center of dilation.

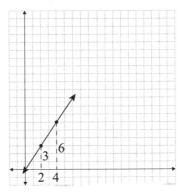

The scale factor for this dilation is 3. Therefore, the dilated version of (2, 3) should be 6 horizontal units and 9 vertical units from the center of dilation.

The dilated version of (4, 6) should be 12 horizontal units and 18 vertical units from the center of dilation.

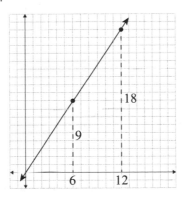

This places the new coordinates at (6, 9) and (12, 18), respectively.

Notice that these new coordinates lie on the original line.

If the center of dilation is a point on the line, then the dilation will leave the line unchanged. In other words, the dilation of any point on that line will always produce a point that already exists on the original line.

Method 2—Arithmetic

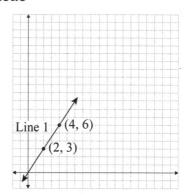

Line 1 $(4, 6)$
$(2, 3)$

Construct a dilated version of line 1 with a scale factor of 1.5 and the center of dilation at (0, 0).

Here is the arithmetic solution for Example 8.

A (2, 3)	B (4, 6)	Original Coordinate
		Note: The center of dilation is (0, 0).
		The scale factor is 3.
2 – 0 = 2 3 – 0 = 3	4 – 0 = 4 6 – 0 = 6	Subtract the coordinates (those of your point minus those of the center of dilation, in that order). (Notice that if the center of dilation is (0, 0), then this step is basically unnecessary.)
2 × 3 = 6 3 × 3 = 9	4 × 3 = 12 6 × 3 = 18	Multiply each of those two numbers by the scale factor.
6 + 0 = 6 9 + 0 = 9	12 + 0 = 12 18 + 0 = 18	Add each of those two numbers, respectively, to the *x*-and *y*-coordinates of the center of dilation. (Notice that if the center of dilation is (0, 0), then this step is basically unnecessary.)
(6, 9)	(12, 18)	Those are the dilated versions of the original coordinates.

Finally, plot the new coordinates and connect them to form a line.

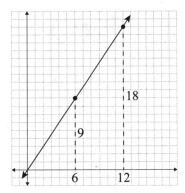

Lesson 3.2
Triangle Similarity

As we learned in Chapter 2, triangles have a certain amount of predictability, which allows us to make some conclusions from limited information. In this lesson, we'll introduce three postulates that you can use to prove triangle similarity. Note that these postulates do *not* work to establish congruence, only similarity.

SSS (SIDE-SIDE-SIDE) SIMILARITY POSTULATE

EXAMPLE 🔒9

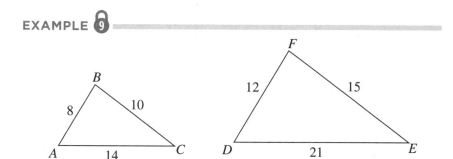

Are triangles *ABC* and *DEF* similar?

In order to prove that these triangles are similar, we can try to prove that there is a constant ratio between the corresponding sides.

Sides *AB* and *DF* correspond, since each is the smallest side of its respective triangle. The ratio of *AB* to *DF* is 8 to 12, or $\frac{8}{12}$. This reduces to $\frac{2}{3}$.

The next largest pair of corresponding sides is *BC* and *FE*. The ratio of *BC* to *FE* is 10 to 15, or $\frac{10}{15}$. This reduces to $\frac{2}{3}$.

The last pair of corresponding sides is *AC* and *DE*. The ratio of *AC* to *DE* is 14 to 21, or $\frac{14}{21}$. This reduces to $\frac{2}{3}$.

All three pairs of corresponding sides are proportional, with a constant ratio of $\frac{2}{3}$.

Therefore, using the definition of similarity, we know that triangles *ABC* and *DEF* are similar.

SSS (Side-Side-Side) Similarity Postulate

Two triangles are similar if all three pairs of corresponding sides are proportional.

EXAMPLE 10

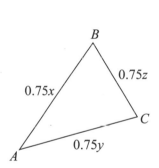

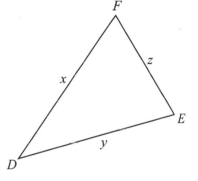

Are triangles *ABC* and *DEF* similar?

This example has just about the same idea as the previous example, but with algebraic expressions instead of numerical values. Here, each pair of corresponding sides shares a variable in common (*x*, *y*, or *z*). You just need to compare the ratios of the corresponding sides. You can easily see that each pair has the same ratio.

Sides *AB* and *DF* correspond, with a ratio of 0.75*x* to 1*x*. This simplifies to $\frac{0.75}{1}$, or $\frac{3}{4}$.

Sides BC and FE correspond, with a ratio of $0.75y$ to $1y$. This simplifies to $\dfrac{0.75}{1}$, or $\dfrac{3}{4}$.

Sides AC and DE correspond, with a ratio of $0.75z$ to $1z$. This simplifies to $\dfrac{0.75}{1}$, or $\dfrac{3}{4}$.

All three pairs of corresponding sides are proportional, with a constant ratio of $\dfrac{3}{4}$. Therefore, using the definition of similarity, we know that triangles ABC and DEF are similar.

SSS Similarity Postulate Exercise

For each of the pairs of triangles below, write "yes" if the triangles can be proved to be similar to the information provided; otherwise, write "no."

Complete the exercise on your own. Answers can be found at the end of the chapter.

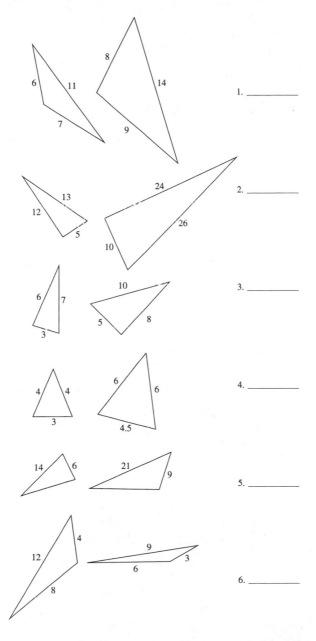

1. _____

2. _____

3. _____

4. _____

5. _____

6. _____

AAA (ANGLE-ANGLE-ANGLE) SIMILARITY POSTULATE

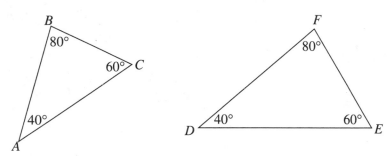

Are triangles *ABC* and *DEF* similar?

One way to show how this postulate works is to make the triangles overlap. Let's place triangle *ABC* within triangle *DEF* so that they share a common angle.

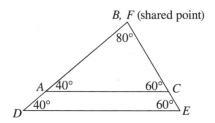

Here, we can prove that segment *AC* is parallel to segment *DE*. That's because the corresponding angles are congruent. (Angle *BAC* corresponds with angle *FDE*, and angle *BCA* corresponds with angle *FED*). Additionally, segments *AB* and *DF* overlap each other (they lie on the same line), as do segments *BC* and *FE*. The triangles are the exact same shape, but they are different sizes. Therefore, they are similar.

EXAMPLE

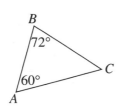

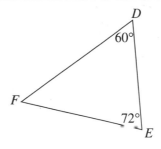

Are triangles *ABC* and *DEF* similar?

In this example, only two pairs of angles are shown. Are the triangles still similar?
Consider what you know about this figure. The sum of angles in any triangle must be
180°. Therefore, you can solve for the missing angle in both triangles.

$\angle A + \angle B + \angle C = 180°$	$\angle D + \angle E + \angle F = 180°$
$60° + 72° + \angle C = 180°$	$60° + 72° + \angle F = 180°$
$\angle C = 180° - (60° + 72°)$	$\angle F = 180° - (60° + 72°)$
$\angle C = 180° - (132°)$	$\angle F = 180° - (132°)$
$\angle C = 48°$	$\angle F = 48°$

Therefore, all of the angles are known, not just two.

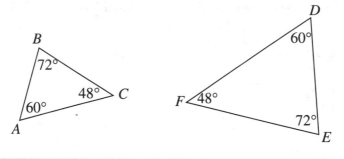

We can apply the AAA Similarity Postulate if we know that we have two pairs of corresponding angles that are congruent.

Here is how you may see the triangle similarity on the ACT.

Two similar triangles have perimeters in the ratio 5:6. The sides of the larger triangle measure 12 in, 7 in, and 5 in. What is the perimeter, in inches, of the smaller triangle?

   F. 18
   G. 20
   H. 22
   J. 24
   K. 32

AAA Similarity Postulate Exercise

For each of the pairs of triangles below, write "yes" if the triangles can be proved to be similar to the information provided; otherwise, write "no."

Complete the exercise on your own. Answers can be found at the end of the chapter.

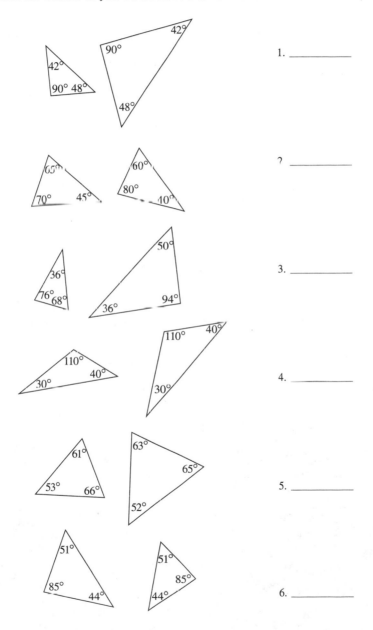

1. _____

2. _____

3. _____

4. _____

5. _____

6. _____

SAS (SIDE-ANGLE-SIDE) SIMILARITY POSTULATE

EXAMPLE 🔒 13

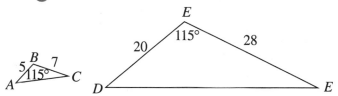

Are triangles *ABC* and *DEF* similar?

In this example, we have two sides given for each triangle, as well as the included angle of each pair. We also know that the two angles are congruent. Next, compare the corresponding sides to see if there is a constant ratio among them.

Sides *AB* and *DE* correspond, since they are each the smallest side of their respective triangles. The ratio of *AB* to *DE* is 5 to 20, or $\frac{5}{20}$. This reduces to $\frac{1}{4}$.

The next largest pair of corresponding sides is *BC* and *EF*. The ratio of *BC* to *EF* is 7 to 28, or $\frac{7}{28}$. This reduces to $\frac{1}{4}$.

The two pairs of corresponding given sides are proportional, with a constant ratio of $\frac{1}{4}$.

The perimeter of the larger triangle is 12 in. + 7 in. + 5 in. = 24 in. Since you know the ratio of the perimeters, you do not have to find each side of the smaller triangle; you can set up the following ratio:

$$\frac{5}{6} = \frac{x}{24}$$

$$\frac{5(24)}{6} = x$$

$x = 20$ in

Therefore, the answer is (G).

How do we know if the third sides are proportional as well?

Let's overlap these triangles, with the common angle as a shared point.

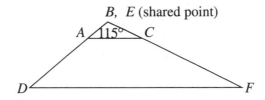

B, E (shared point)

A 115° *C*

D　　　　　　　*F*

With the figures overlapping, it's easy to see that sides *AC* and *DF* appear to be

parallel. In fact, we can prove that they are parallel because the given pairs of

corresponding sides have a common ratio of $\frac{1}{4}$. That is, points *A* and *C* are both $\frac{1}{4}$ of

the distance away from the shared vertex, and that means that the slope of *AC* is the

same as that of *DF*.

Since *AC* and *DF* are parallel, we know that the corresponding angles *BAC* and *EDF*
are congruent, as well as the corresponding angles *ACB* and *DFE*. Therefore, all three
pairs of corresponding angles are congruent, and using the definition of similarity,
we know that triangles *ABC* and *DEF* are similar.

SAS (Side-Angle-Side) Similarity Postulate

Two triangles are similar if two pairs of corresponding sides
are proportional and the included angles are congruent.

SAS Similarity Postulate Exercise

For each of the pairs of triangles below, write "yes" if the triangles can be proved to be similar with the information provided; otherwise, write "no."

Complete the exercise on your own. Answers can be found at the end of the chapter.

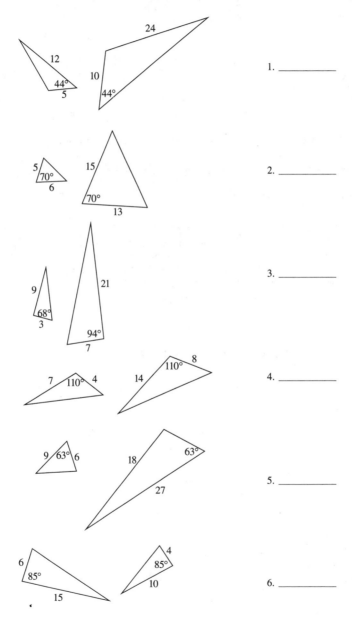

1. _____

2. _____

3. _____

4. _____

5. _____

6. _____

ANSWERS TO CHAPTER 3 EXERCISES

SSS Similarity Postulate Exercise (Page 131)

1. No.
   These triangles do not have proportional sides; the ratios of corresponding sides are 6:8 (or 3:4), 7:9, and 11:14, respectively. Therefore, they are not similar.

2. Yes.
   These triangles have the same proportions for all three pairs of corresponding sides; therefore, they are similar.

3. No.
   These triangles have two pairs of proportional sides (both 3:5 and 6:10 reduce to 3:5). However, the third pair of corresponding sides is not in the same proportion (7:8). Therefore, they are not similar.

4. Yes.
   These triangles have the same proportions for all three pairs of corresponding sides; therefore, they are similar.

5. No.
   These triangles have two pairs of proportional sides (both 6:14 and 9:21 reduce to 3:7). However, one side length is unknown; therefore, they are not necessarily similar.

6. Yes.
   These triangles have the same proportions for all three pairs of corresponding sides; therefore, they are similar.

AAA Similarity Postulate Exercise (Page 135)

1. Yes.
   These triangles have three pairs of congruent corresponding angles; therefore, they are similar.

2. No.
   These triangles do not have congruent corresponding angles. Therefore, they are not similar.

3. No.
   These triangles have only one pair of corresponding congruent angles. However, the other pairs are not congruent. Therefore, the triangles are not similar.

4. Yes.
These triangles have three pairs of congruent corresponding angles; therefore, they are similar.

5. No.
These triangles have pairs of angles that are *close* in measure, but the angles are not congruent. We cannot apply AAA if the angles are not exactly congruent. Therefore, these triangles are not similar.

6. Yes.
These triangles each have two known angles, which means that the third angle can be derived for each as well. These triangles have three pairs of congruent corresponding angles; therefore, they are similar.

SAS Similarity Postulate Exercise (Page 138)

1. Yes.
These triangles have the same proportions for both pairs of corresponding sides, and the included angles are congruent; therefore, they are similar.

2. No.
These triangles do not have proportional sides; the ratios of corresponding sides are 5:15 (or 1:3) and 6:13, respectively. Therefore, they are not similar.

3. No.
These triangles have two pairs of proportional sides (both 3:7 and 9:21 reduce to 3:7). However, the included angles are not congruent. Therefore, they are not similar.

4. Yes.
These triangles have the same proportions for both pairs of corresponding sides, and the included angles are congruent; therefore, they are similar.

5. No.
These triangles have two pairs of proportional sides (both 6:18 and 9:27 reduce to 1:3). However, the given angles do not correspond (they are in a different place in the two triangles). Therefore, the triangles are not similar.

6. Yes.
These triangles have the same proportions for both pairs of corresponding sides, and the included angles are congruent; therefore, they are similar.

CHAPTER 3 PRACTICE QUESTIONS

Directions: Complete the following problems as specified by each question. For extra practice after answering each question, try using an alternative method to solve the problem or check your work. Larger, printable versions of images are available in your online student tools.

1. Construct a dilated version of triangle *XYZ* with a scale factor of 2.5 and the center of dilation at vertex *Z*.

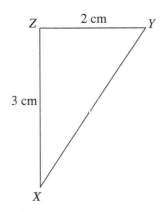

3. Construct a dilated version of quadrilateral *HIJK*, shown below, with a scale factor of 2 and the center of dilation at (0, 0).

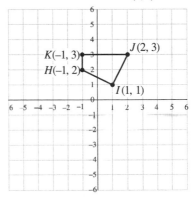

2. In the figure below, hexagon *MNOPQR* is dilated with a scale factor of 0.5. Find the center of dilation.

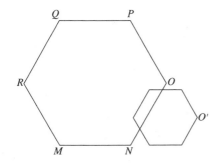

4. Construct a dilated version of quadrilateral *HIJK*, shown below, with a scale factor of 2 and the center of dilation at (−1, −1).

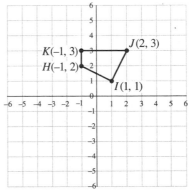

5. Construct a dilated version of the line $y = 2x + 5$ with a scale factor of 0.5 and a center of dilation at $(0, 0)$. Show both the original line and the dilation on the graph. Determine the equation of the dilated line.

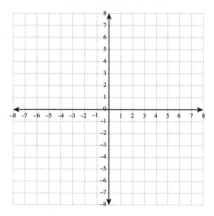

6. A contractor has drawn a rectangular kitchen on a coordinate grid using the points $(5, 9)$, $(5, 21)$, $(-1, 9)$, and $(-1, 21)$. The client, however, would like each dimension of the kitchen to be 1.5 times the current size. The point $(-1, 9)$ will be kept the same. What will be the other three points?

7. A jogger travels 3 miles due west from her home to the park, 4 miles due north from the park to the lake, and 5 miles from the lake directly back home every morning. One morning she decides to extend her jog and travel one additional mile due west and one additional mile due north before jogging home. Do the two routes represent similar triangles? Explain.

8. If possible, calculate the length of side *EF* without using trigonometry.

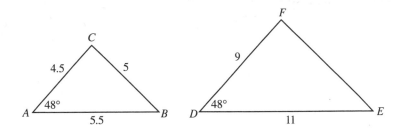

9. If possible, calculate the length of side *KL* without using trigonometry.

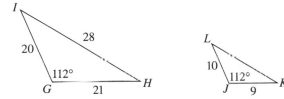

SOLUTIONS TO CHAPTER 3 PRACTICE QUESTIONS

1.

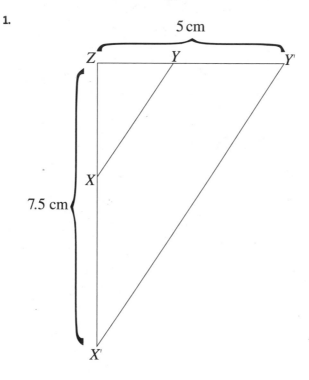

The scale factor is 2.5. Multiply the length of each segment by 2.5 to determine the lengths for the dilated triangle. Segment *ZX'* will be 7.5 cm (3 × 2.5), and segment *ZY'* will be 5 cm (2 × 2.5). Next, draw these new segments: extend segment *ZX* to a length of 7.5 cm and segment *ZY* to a length of 5 cm. Since *Z* is the center of dilation, this is the originating point for your measurements. Finally, connect segment *X'Y'* to have the full dilated triangle. You can also verify that segment *X'Y'* is 2.5 times the length of segment *XY* (this is optional).

2.

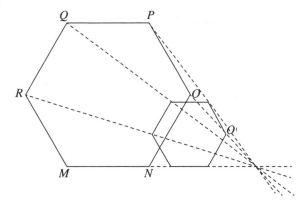

For each point on *MNOPQR*, draw a line through that point and its corresponding point on the dilation. These lines should all intersect at a single point, which is the center of dilation. Note: You can find the intersection point with just two of these lines! But the more lines you draw, the more likely that you can avoid mistakes and be more confident in your answer.

Note: If you do draw all of these lines, it can be a bit difficult to get the intersection point to be 100% precise. This is most likely due to understandable human error with your pencil and ruler. However, you should keep practicing dilations to get them as precise as possible!

3.

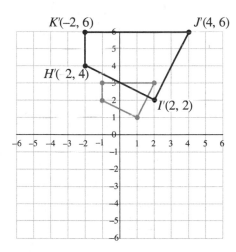

Since the center of dilation is at the origin, multiply each coordinate by the scale factor of 2. The new vertices are (−2, 4), (2, 2), (4, 6), and (−2, 6).

4.

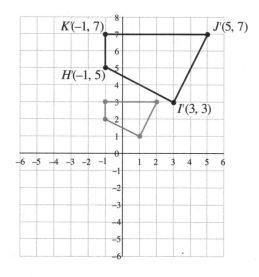

Follow the steps discussed in the lesson:

H (−1, 2)	I (1, 1)	J (2, 3)	K (−1, 3)	Original Coordinate Note: The center of dilation is (−1, −1). The scale factor is 2.
−1 − (−1) = 0 2 − (−1) = 3	1 − (−1) = 2 1 − (−1) = 2	2 − (−1) = 3 3 − (−1) = 4	−1 − (−1) = 0 3 − (−1) = 4	Subtract the coordinates (those of your point minus those of the center of dilation, in that order).
0 × 2 = 0 3 × 2 = 6	2 × 2 = 4 2 × 2 = 4	3 × 2 = 6 4 × 2 = 8	0 × 2 = 0 4 × 2 = 8	Multiply each of those two numbers by the scale factor.
(−1) + 0 = −1 (−1) + 6 = 5	(−1) + 4 = 3 (−1) + 4 = 3	(−1) + 6 = 5 (−1) + 8 = 7	(−1) + 0 = −1 (−1) + 8 = 7	Add each of those two numbers, respectively, to x- and y- coordinates of the center of dilation.
(−1, 5)	(3, 3)	(5, 7)	(−1, 7)	Those are the dilated versions of the original coordinates.

5.

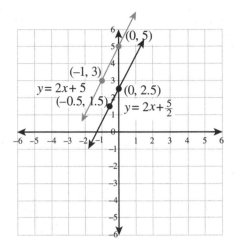

Start by finding two points on the line $y = 2x + 5$. One easy point is the y-intercept, where $x = 0$. When $x = 0$, $y = 5$. One point is (0, 5). Now find another point. Try putting in −1 for x. In that case, $y = 3$. A second point is (−1, 3). Now draw both points on the graph and connect them to draw the complete line. Since the center of dilation is at the origin, each point just needs to be multiplied by the scale factor, 0.5. The new points are (0, 2.5) and (−0.5, 1.5). Draw those points and connect them to form the line. Last, find the equation of that line. Since we have the point (0, 2.5), we know the y-intercept is 2.5 or 5/2. The slope is the same as on the original line since dilated lines are parallel. Thus, the formula is $y = 2x + 5/2$.

6. **(8, 9), (8, 27), (−1, 27)**

Translate this into a math problem. This is a dilation with a scale factor of 1.5, and the center of dilation at point (−1, 9). Start with the point (5, 9). It is 6 spaces away from (−1, 9), so multiply 6 by 1.5 to get 9. The new length of that side of the kitchen is 9, so go 9 spaces to the right from (−1, 9) to get that the new point, which will be (8, 9).

Now try (−1, 21). It is 12 spaces away from the center of dilation, (−1, 9), so multiply 12 by 1.5 to get 18. The new distance is 18 spaces up from (−1, 9), so that is (−1, 27).

For the last point, the easiest way may be to refer to the previous two points. Since it's a rectangle, the x-value must be 8 and the y-value must be 27, since this corresponds to the adjacent vertices in the figure. The third coordinate is (8, 27).

7. **No**

   Draw the two right triangles. One has legs of 3 and 4 and the other has legs of 4 and 5. Although the 90 degree angle between them is the same, the two pairs of sides are not proportional because 4/5 is not equal to 3/4. Since the corresponding sides are not proportional, the two triangles are not similar.

8. **10**

   Notice that the two triangles have at least one congruent angle. Compare the corresponding sides. 4.5/9 = 1/2 and 5.5/11 = 1/2, so the two sides given have the same ratio. Thus, the two triangles are similar, by the SAS similarity postulate. Side *FE* must be twice side *CB*, so it is equal to 10.

9. **Not possible**

   The two triangles have at least one congruent angle. Examine the corresponding sides. 20 is twice 10, but 21 is not twice 9. This means the two sides provided do not have the same ratio. The triangles are not similar and therefore it is not possible to calculate the missing side without using trigonometry.

REFLECT

Congratulations on completing Chapter 3! Here's what we just covered. Rate your confidence in your ability to:

- Understand and perform dilations of figures, including figures in the coordinate plane

  ① ② ③ ④ ⑤

- Identify similar triangles using the SSS, AAA, and SAS similarity postulates

  ① ② ③ ④ ⑤

If you rated any of these topics lower than you'd like, consider reviewing the corresponding lesson before moving on, especially if you found yourself unable to correctly answer one of the related end-of-chapter questions.

 Access your online student tools for a handy, printable list of Key Points for this chapter. These can be helpful for retaining what you've learned as you continue to explore these topics.

Chapter 4
Trigonometry

4

GOALS By the end of this chapter,
you will be able to:

- Understand and apply the basic trigonometric functions (sine, cosine, and tangent) and their reciprocals (secant, cosecant, and cotangent)

- Use trigonometric functions and their inverses on your calculator

- Find sine, cosine, and tangent values for complementary angles

- Use trigonometric functions to solve problems with right triangles

- Use the Pythagorean theorem to derive and understand additional trigonometric identities

- Use the Law of Sines and Law of Cosines to solve problems with non-right triangles

Lesson 4.1
Trigonometric Ratios

WHAT IS TRIGONOMETRY?

Trigonometry is the study of triangles. It's actually so important that there is an entire branch of mathematics devoted to this topic. The most common ways that you'll use trigonometry in high school include solving for unknown side lengths and/or angles in triangles. In the real world, trigonometry has many applications in other fields such as physics, engineering, astronomy, and even music.

Here is how you may see trigonometric ratios on the SAT.

In rectangle *PQRS*, shown below, the diagonal *PR* is 15 meters. If the sine of ∠*SPR* is $\frac{7}{10}$, what is the value of *RS* ?

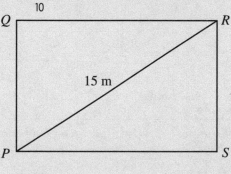

A) 0.01
B) 0.70
C) 7.0
D) 10.5

TRIGONOMETRIC RATIOS

Almost everything we know about trigonometry can be derived from relationships found in right triangles. In fact, some basics that you already know—like the Pythagorean theorem—play a fundamental role. Trigonometry allows us to use the known **ratios** of a triangle to solve for unknown information, like side lengths.

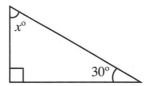

Looking at the triangle above, you know that one angle is 90° and one angle is 30°. What else do you know? Since the angles in a triangle always add up to 180°, you'd also be able to find x. 180° − (90° + 30°) = 60°, so x would be 60°. You have all three angles for this triangle, then, which also means you know its proportions. That's the fundamental theorem of trigonometry—if you know the angles of a triangle, you know the **ratios** of the sides. That is, you might not know the actual values of the individual sides, but you know the relationships between them.

Based on what we know of triangle proportions, trigonometry has three basic functions to express these relationships.

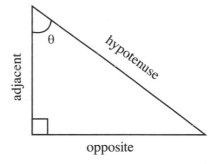

Here, we'll define some terms. Then, you'll see some examples.

θ (called "theta")—a commonly used variable for angles. (Think of it like an *x*.)
Opposite and *adjacent*—the two **legs** of the triangle.
The *opposite* side—the leg that's opposite from (i.e. not touching) the angle θ.
The *adjacent* side—the leg that's adjacent to (i.e. touching) the angle θ.
The *hypotenuse*—the longest side of a right triangle. It's always opposite from the 90° angle.

Sine

The sine of an angle is the ratio of its opposite side to the hypotenuse.
The sine function of an angle θ is abbreviated as sin θ.

Cosine

The cosine of an angle is the ratio of its adjacent side to the hypotenuse.
The cosine function of an angle θ is abbreviated as cos θ.

Tangent

The tangent of an angle is the ratio of its opposite side to its adjacent side.
The tangent function of an angle θ is abbreviated as tan θ.

If the sine of ∠*SPR* is $\frac{7}{10}$, that would mean the diagonal of the rectangle is 10.
We know that it's 15, not 10, so we need to set up a proportion. Your proportion

should look like this: $\frac{7}{10} = \frac{x}{15}$. When you solve for *x*, you get 10.5 meters.

The correct answer is (D).

A mnemonic for these functions is SOHCAHTOA (pronounce it like "so-ca-toe-a").

$$S = \frac{O}{H}$$

$$C = \frac{A}{H}$$

$$T = \frac{A}{O}$$

SOHCAHTOA

$$\mathbf{s}in = \frac{\mathbf{o}pposite}{\mathbf{h}ypotenuse} \qquad \mathbf{c}os = \frac{\mathbf{a}djacent}{\mathbf{h}ypotenuse} \qquad \mathbf{t}an = \frac{\mathbf{o}pposite}{\mathbf{a}djacent}$$

Let's take another look at the 30°-60°-90° triangle. Remember that this is a "special" right triangle, whose proportions you may have memorized previously.

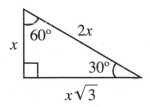

For example, if the short side is 2, then the triangle would have the following side lengths.

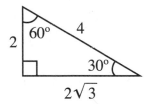

Here are the trigonometric ratios for these angles.

	30°	60°
Sine (*opposite/hypotenuse*)	$= \dfrac{2}{4}$ $= \dfrac{1}{2}$ $= 0.5$	$= \dfrac{2\sqrt{3}}{4}$ $= \dfrac{\sqrt{3}}{2}$ ≈ 0.866
Cosine (*adjacent/hypotenuse*)	$= \dfrac{2\sqrt{3}}{4}$ $= \dfrac{\sqrt{3}}{2}$ ≈ 0.866	$= \dfrac{2}{4}$ $= \dfrac{1}{2}$ $= 0.5$
Tangent (*opposite/adjacent*)	$= \dfrac{2}{2\sqrt{3}}$ $= \dfrac{1}{\sqrt{3}}$ (or, $\dfrac{\sqrt{3}}{3}$) ≈ 0.577	$= \dfrac{2\sqrt{3}}{2}$ $= \sqrt{3}$ ≈ 1.732

Remember, "opposite" and "adjacent" are relative to the angle you're working with. The "adjacent" of 30° is the same as the "opposite" of 60°, and vice versa.

Fractions in simplified form do not have irrational numbers in the denominator. To simplify, multiply the irrational number on the top and bottom of the fraction.

Example: $\dfrac{1}{\sqrt{3}}$

$= \dfrac{1 \times \sqrt{3}}{\sqrt{3} \times \sqrt{3}}$

$= \dfrac{\sqrt{3}}{3}$

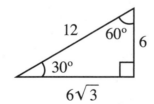

This triangle is larger than the one in the previous example. However, it is still a 30°-60°-90° triangle. In other words, the two triangles are **similar**—they have the same angles, even though they are different sizes.

Using these side lengths, we can see that the trigonometric ratios are still the same.

$$\sin 30° = \frac{6}{12} = \frac{1}{2}$$

$$\cos 30° = \frac{6\sqrt{3}}{12} = \frac{\sqrt{3}}{2}$$

$$\tan 30° = \frac{6}{6\sqrt{3}} = \frac{1}{\sqrt{3}}$$

And so on. Therefore, since similar triangles have the same angles, they also have the same trigonometric relationships.

> Each angle has given values for sine, cosine, and tangent.

Recall the AAA Similarily Postulate from Lesson 3.2.

EXAMPLE 1

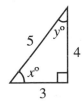

Here's our favorite right triangle, the 3-4-5. Can you determine the sin, cos, and tan values for $x°$ and $y°$?

	$x°$	$y°$
Sin		
Cos		
Tan		

Check your answers at the end of the chapter.

Lesson 4.2
Complementary Angles

In geometry, two angles are **supplementary** to each other if their sum equals 180°. For example, 45° is supplementary to 135°.

When two angles have a sum of 90°, they are known as **complementary** angles. For example, 30° is complementary to 60°. In a right triangle, the two acute angles are always complementary.

EXAMPLE

For each of the angle measures in the table below, find the measure of the complementary angle.

Angle	Complement
20°	
43°	
58°	

Check your answers at the end of the chapter.

The concept of complementary angles is very important in trigonometry. You may have noticed in the examples of the right triangles in the previous section, that the complementary angles have the same sin and cos values—except that they're switched. That is, the sine of one angle is equal to the cosine of its complement. For example, the sine of 30° is equal to the cosine of 60°. (Both are equal to 0.5.)

Additionally, the tangents of complementary angles are **reciprocals** of each other. In other words, the tangent of one angle is equal to the reciprocal of the tangent of its complement. For example, the tangent of 60° is $\sqrt{3}$, while the tangent of 30° is $\frac{1}{\sqrt{3}}$.

Complementary Angles

Complementary angles have a sum of 90°.

The sine of $x°$ is equal to the cosine of $90 - x°$.

The cosine of $x°$ is equal to the sine of $90 - x°$.

The tangent of $x°$ is the reciprocal of the tangent of $90 - x°$.

EXAMPLE

In the table below, use the given value on the left to determine the indicated value on the right.

$\sin(20°) \approx 0.342$	$\cos(70°) \approx$
$\cos(40°) \approx 0.766$	$\sin(50°) \approx$
$\sin(x°) = \dfrac{2}{5}$	$\cos(90 - x°) =$
$\cos(90 - x°) = \dfrac{8}{9}$	$\sin(x°) =$
$\tan(x°) = \dfrac{3}{4}$	$\tan(90 - x°) =$
$\tan(90 - x°) = \dfrac{2}{3}$	$\tan(x°) =$

Check your answers at the end of the chapter.

Your Calculator

Scientific and graphing calculators have sin, cos, and tan built in! When given an angle, the calculator can tell you the value of sin, cos, or tan for that angle.

It is not typically practical to solve inverse functions without a calculator. The inverse function of some "special" right triangles (such as 30°-60°-90° or 45°-45°-90°) can be memorized. One other way to solve inverse functions is to observe a graph of sin, cos, or tan values.

Want to try it? With your calculator in degree mode, enter sin(30°). The calculator should show $\frac{1}{2}$, or 0.5.

You can try the different functions for 30° and 60°, and compare the results with the values in the table shown previously.

Your calculator can also do the *inverse* of these functions, which takes the *ratio* and solves for the *angle.* For example, the inverse of sin30° is just 30°. On your calculator, the inverse of sin might look like sin^{-1}, or "arcsin."

If you enter sin^{-1}(0.5), in degree mode, your calculator should return 30°.

EXAMPLE 4 ═══════════════════════════════════

Which function should you use to solve?

If you have all three side lengths of the triangle, then all three functions work equally well!

What is the value of $x°$ and $y°$? Use sin^{-1}, cos^{-1}, or tan^{-1}.

Check your answers at the end of the chapter.

═══ 4

Lesson 4.3
Problem Solving

EXAMPLE 5

A contractor builds a ramp to reach a loading dock that is 10 feet high. The ramp measures 26 feet along its slanted surface. What is the angle of incline of the ramp?

The ramp forms a right triangle, in which the *opposite* side is 10 feet and the

hypotenuse is 26 feet. Since those are the two given sides, the most straightforward

way to solve this problem is to calculate $\sin^{-1}\left(\dfrac{10}{26}\right)$, or simplify as $\sin^{-1}\left(\dfrac{5}{13}\right)$. Use

your calculator.

$$\sin^{-1}\left(\frac{5}{13}\right) \approx 22.62°$$

5

Note that you can also solve for the unknown side of this triangle, using the Pythagorean theorem.

$$10^2 + b^2 = 26^2$$
$$100 + b^2 = 676$$
$$b^2 = 676 - 100$$
$$b^2 = 576$$
$$b = 24$$

Knowing all three sides of the triangle, you can now use the other two inverse identities to solve for the same angle.

$$\sin^{-1}\left(\frac{5}{13}\right) \approx 22.62°$$

$$\cos^{-1}\left(\frac{12}{13}\right) \approx 22.62°$$

$$\tan^{-1}\left(\frac{5}{12}\right) \approx 22.62°$$

Did we *need* to do that here? Of course not. However, it's helpful to know your options for more difficult problems. Some exercises in school, or on standardized tests, may challenge you to use specific functions, instead of the most straightforward one.

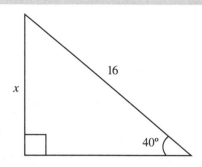

What is the value of *x*?

This is quite typical of the type of trig problem you might see in school, or on standardized tests. You have only one given side length, so you can't use the Pythagorean theorem. However, you do have an angle, which allows you to use sin, cos, or tan.

To decide which function to use, consider which side length you have, and which one you need. In this case, you have the hypotenuse (16), and you need the opposite side. So, use sin(40°). Here's how.

$\sin(x) = \dfrac{\text{opposite}}{\text{hypotenuse}}$	Start with the basic equation for sin.
$\sin(40°) = \dfrac{x}{16}$	Enter your known values in the equation.
$16 \times \sin(40°) = x$ $x \approx 16 \times 0.6428$ $x \approx 10.285$	Simplify. Use your calculator.

OTHER TRIGONOMETRIC IDENTITIES

Reciprocal Identities

The reciprocals of sine, cosine, and tangent are, respectively, **cosecant** (abbreviated csc), **secant** (abbreviated sec), and **cotangent** (abbreviated cot). This builds on the relationships of sin, cos, and tan, and can come in handy with more advanced problems. For now, just work on memorizing these relationships.

$$\text{csc} = \frac{\text{hypotenuse}}{\text{opposite}} = \frac{1}{\text{sin}}$$

$$\text{sec} = \frac{\text{hypotenuse}}{\text{adjacent}} = \frac{1}{\text{cos}}$$

$$\text{cot} = \frac{\text{adjacent}}{\text{opposite}} = \frac{1}{\text{tan}}$$

Observe the ratios in the triangle below.

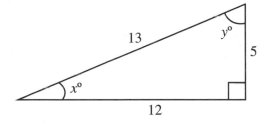

	$x°$	$y°$
Sine (*opposite/hypotenuse*)	$=\dfrac{5}{13}$ ≈ 0.385	$=\dfrac{12}{13}$ ≈ 0.923
Cosine (*adjacent/hypotenuse*)	$=\dfrac{12}{13}$ ≈ 0.923	$=\dfrac{5}{13}$ ≈ 0.385
Tangent (*opposite/adjacent*)	$=\dfrac{5}{12}$ ≈ 0.417	$=\dfrac{12}{5}$ $=2.4$
Cosecant (*hypotenuse/opposite*)	$=\dfrac{13}{5}$ $=2.6$	$=\dfrac{13}{12}$ ≈ 1.083
Secant (*hypotenuse/adjacent*)	$=\dfrac{13}{12}$ ≈ 1.083	$=\dfrac{13}{5}$ $=2.6$
Cotangent (*adjacent/opposite*)	$=\dfrac{12}{5}$ $=2.4$	$=\dfrac{5}{12}$ ≈ 0.417

ACT

Here is how you may see reciprocal identities on the ACT.

If $g(x) = \sin x \cot x$, then which of the following trigonometric functions is equivalent to $g(x)$?

(Note: $\csc x = \dfrac{1}{\sin x}$, $\sec x = \dfrac{1}{\cos x}$, and $\cot x = \dfrac{1}{\tan x}$)

A. $g(x) = \sin x$
B. $g(x) = \cos x$
C. $g(x) = \tan x$
D. $g(x) = \csc x$
E. $g(x) = \sec x$

The Pythagorean Identity

The Pythagorean Identity is based on—you guessed it—the Pythagorean theorem. Let's see how this is derived.

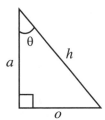

For all angles, the value of sin and cos is always between –1 and 1 (inclusive).

Tan does not have this limit.

Remember: sine of $x°$ is written as sin($x°$).

Theta (θ) is a variable for an angle.

$$\sin^2(\theta) + \cos^2(\theta) = 1$$

$$o^2 + a^2 = h^2$$

Start with the Pythagorean theorem. Since this is a right triangle, we know the theorem will be satisfied.

An **identity** is an equation or formula that's true for all values.

$$\sin^2(\theta) + \cos^2(\theta)$$

$$= \frac{o^2}{h^2} + \frac{a^2}{h^2}$$

$$= \frac{o^2 + a^2}{h^2}$$

If we square sin(θ), we get $\dfrac{o^2}{h^2}$.

If we square cos(θ), we get $\dfrac{a^2}{h^2}$.

$$= \frac{o^2 + a^2}{o^2 + a^2}$$

$$= \frac{h^2}{h^2} = 1$$

Using the Pythagorean theorem, we know that the numerator and denominator are equal. ($o^2 + a^2 = h^2$)

See this in action with a 3-4-5 triangle:

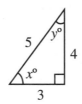

$$\sin(x) = \frac{3}{5}$$

$$\cos(x) = \frac{4}{5}$$

$$\sin^2(x) = \left(\frac{4}{5}\right)^2 = \frac{16}{25}$$

$$\cos^2(x) = \left(\frac{3}{5}\right)^2 = \frac{9}{25}$$

$$\sin^2(\theta) + \cos^2(\theta) = \frac{16}{25} + \frac{9}{25} = \frac{25}{25} = 1$$

$$\sin = \frac{opp}{hyp}$$

$$\cot = \frac{1}{\tan} = \frac{adj}{opp}$$

$$\sin \times \cot$$

$$= \frac{opp}{hyp} \times \frac{adj}{opp}$$

$$= \frac{adj}{hyp} \qquad \text{Simplify}$$

$$= \cos$$

The correct answer is (B).

Here's another way to think about this identity.

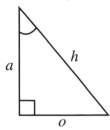

$o^2 + a^2 = h^2$	Start with the Pythagorean theorem. Since this is a right triangle, we know the theorem will be satisfied.
$\dfrac{o^2}{h^2} + \dfrac{a^2}{h^2} = \dfrac{h^2}{h^2}$ $\dfrac{o^2}{h^2} + \dfrac{a^2}{h^2} = 1$	Divide both sides by h^2.
$\sin^2(\theta) + \cos^2(\theta) = 1$	Observe that $\dfrac{o^2}{h^2} = \left(\dfrac{o}{h}\right)^2 = \sin^2(\theta)$. Observe that $\dfrac{a^2}{h^2} = \left(\dfrac{a}{h}\right)^2 = \cos^2(\theta)$.

From this identity, we can derive all of the following:

$$\sin^2(\theta) + \cos^2(\theta) = 1$$
$$1 - \sin^2(\theta) = \cos^2(\theta)$$
$$1 - \cos^2(\theta) = \sin^2(\theta)$$
$$\tan^2(\theta) + 1 = \sec^2(\theta)$$
$$\cot^2(\theta) + 1 = \csc^2(\theta)$$

...and more!

Sin/Cos

Another very useful identity is the following:

$$\frac{\sin(\theta)}{\cos(\theta)} = \tan(\theta)$$

Let's see how this identity is derived.

$\dfrac{\sin(\theta)}{\cos(\theta)}$	$\sin(\theta) = \dfrac{o}{h}$
	$\cos(\theta) = \dfrac{a}{h}$

$= \dfrac{\dfrac{o}{h}}{\dfrac{a}{h}}$

$= \dfrac{o}{h} \times \dfrac{h}{a}$ To divide, multiply by the reciprocal.

$= \dfrac{oh}{ha}$ Cancel h from the numerator and denominator.

$= \dfrac{o}{a}$ $\tan(\theta) = \dfrac{o}{a}$

$= \tan(\theta)$

Let's see this identity with a 5-12-13 triangle.

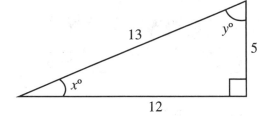

To see how trigonometric identities are tested on the ACT, access your Student Tools online.

$\sin(x°)/\cos(x°)$

$$= \frac{\dfrac{5}{13}}{\dfrac{12}{13}}$$

$$= \frac{5}{13} \times \frac{13}{12}$$

$$= \frac{5 \times 13}{13 \times 12}$$

$$= \frac{5}{12}$$

$$= \tan(x°)$$

From this identity, we can derive the following:

$$\frac{\sin(\theta)}{\cos(\theta)} = \tan(\theta)$$

$$\frac{\cos(\theta)}{\sin(\theta)} = \cot(\theta)$$

...and more!

Lesson 4.4
Trigonometry with Non-Right Triangles

Sometimes, we know more than we think we do...

EXAMPLE 7

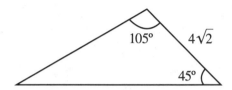

What is the area of the triangle above?

In order to find the area, you'll need to find the lengths of the height and base. First, draw a line representing the height. Label the unknown sides as *a*, *b*, and *c*. The height must always be perpendicular to the base, so you know it forms a right angle. Therefore, the smaller triangle with 45° and 90° must be a 45°-45°-90° triangle.

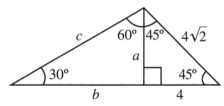

You can use the given side length of $4\sqrt{2}$ to find more information. $4\sqrt{2}$ is the hypotenuse of our 45°-45°-90° triangle. Therefore, the two legs of this smaller triangle are each equal to 4.

To prove this, recall the "special" right triangle 45°-45°-90° has side lengths *x*, *x*, and $x\sqrt{2}$.

Or, you can use the Pythagorean theorem: $a^2 + a^2 = \left(4\sqrt{2}\right)^2$. (The 45°-45°-90° triangle is isosceles).

Or, you can use $\sin(45°) = \dfrac{a}{4\sqrt{2}}$.

That's a lot of options!

height = 4

Now that you know the height of this triangle, you can also find the other unknown side lengths.

If the height is 4, then b is $4\sqrt{3}$, and c must be 8.

To prove this, use the relationships of the "special" right triangle 30°-60°-90°.

Or, use sin, cos, or tan functions to solve. (For example, use $\tan(60°) = \dfrac{b}{4}$ and $\cos(60°) = \dfrac{c}{4}$).

base $= 4\sqrt{3} + 4$

≈ 10.93

Therefore, the area of the triangle is determined as follows:

$A = \dfrac{1}{2}bh$

$\approx \dfrac{1}{2} \times 4 \times (10.93)$

$\approx 2 \times (10.93)$

≈ 21.86

The Law of Sines

Because of the ability to make triangles within triangles, yet another identity can be derived.

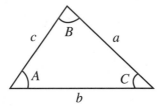

The Law of Sines

$$\frac{\sin(A)}{a} = \frac{\sin(B)}{b} = \frac{\sin(C)}{c}$$

in which side a is opposite angle A, side b is opposite angle B, and side c is opposite angle C.

The way to use the Law of Sines is to create a proportion. Each angle corresponds with its opposite side.

Let's take another look at the 105-30-45 triangle.

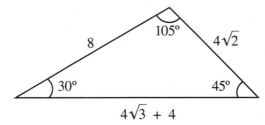

Using the Law of Sines, we have the following proportion:

$$\frac{\sin(30°)}{4\sqrt{2}} = \frac{\sin(45°)}{8} = \frac{\sin(105°)}{4\sqrt{3}+4}$$

Use your calculator to confirm that the proportion is true:

$$\frac{\sin(30°)}{4\sqrt{2}} \approx 0.0884$$

$$\frac{\sin(45°)}{8} \approx 0.0884$$

$$\frac{\sin(105°)}{4\sqrt{3}+4} \approx 0.0884$$

We can use the Law of Sines to solve for unknown side lengths or angles.

EXAMPLE 🔒

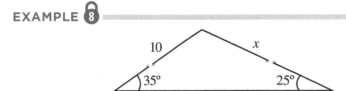

What is the value of x?

To solve, use the Law of Sines. Note that in most cases, a proportion can be two fractions instead of three. You don't always need all three "sides" to the Law of Sines equation.

$\dfrac{\sin(A)}{a} = \dfrac{\sin(B)}{b} = \dfrac{\sin(C)}{c}$	Use the Law of Sines.
$\dfrac{\sin(25°)}{10} = \dfrac{\sin(35°)}{x}$	Correctly plug the given information into the equation. The side labeled 10 corresponds with sin(25°).
$10 \times \sin(35°) = x\,(\sin(25°))$	Cross-multiply.
$10 \times \dfrac{\sin(35°)}{\sin(25°)} = x$	Divide by sin(25°).
$10 \times \dfrac{0.574}{0.423}$	Use your calculator.
≈ 13.57	

🔓 8

The Law of Cosines

The Law of Cosines is based on three sides and one angle. It allows you to solve if one of those facts is unknown.

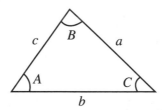

The Law of Cosines

$$c^2 = a^2 + b^2 - 2ab \, (\cos(C))$$

in which side a is opposite angle A, side b is opposite angle B, and side c is opposite angle C.

Here is how you may see the Law of Sines and the Law of Cosines on the ACT.

Triangle JKL is shown in the figure below. The measure of $\angle K$ is 50°, $JK = 9$ cm, and $KL = 6$ cm. Which of the following is the length, in centimeters, of LJ?

Note: For a triangle with sides of length a, b, and c opposite angles $\angle A$, $\angle B$, and $\angle C$, respectively, the Law of Sines states $\dfrac{\sin \angle A}{a} = \dfrac{\sin \angle B}{b} = \dfrac{\sin \angle C}{c}$, and the Law of Cosines states

$c^2 = a^2 + b^2 - 2ab \, (\cos(C))$.

A. $9\sin 50°$

B. $6\sin 50°$

C. $\sqrt{9^2 - 6^2}$

D. $\sqrt{9^2 + 6^2}$

E. $\sqrt{9^2 + 6^2 - 2(9)(6)\cos 50°}$

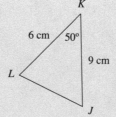

See it in action with a 30°-60°-90° triangle.

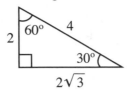

The most important thing to remember when setting up the Law of Cosines, is that the angle C is opposite side c.

There are a few different ways we can set up the Law of Cosines with this figure.

$$2^2 = 4^2 + \left(2\sqrt{3}\right)^2 - 2(4 \times 2\sqrt{3})(\cos(30°))$$

$$\left(2\sqrt{3}\right)^2 = 4^2 + 2^2 - 2(4 \times 2)(\cos(60°))$$

$$4^2 = 2^2 + \left(2\sqrt{3}\right)^2 - 2(2 \times 2\sqrt{3})(\cos(90°))$$

There are no unknowns in this example, so just simplify and use your calculator to see that the equations are true. We'll do the first one step-by-step:

$$2^2 = 4^2 + \left(2\sqrt{3}\right)^2 - 2(4 \times 2\sqrt{3})(\cos(30°))$$
$$4 = 4^2 + \left(2\sqrt{3}\right)^2 - 2(4 \times 2\sqrt{3})(\cos(30°))$$
$$4 = 16 + \left(2\sqrt{3}\right)^2 - 2(4 \times 2\sqrt{3})(\cos(30°))$$
$$4 = 16 + 12 - 2(4 \times 2\sqrt{3})(\cos(30°))$$
$$4 = 16 + 12 - 2(8\sqrt{3})(\cos(30°))$$
$$4 = 16 + 12 - (16\sqrt{3})(\cos(30°))$$
$$4 \approx 16 + 12 - (27.71)(\cos(30°))$$
$$4 \approx 16 + 12 - (27.713)(0.866)$$
$$4 \approx 16 + 12 - (23.99)$$
$$4 \approx 28 - (23.99)$$
$$4 \approx 4.01$$

That's about right! If you rounded the irrational numbers, you can expect the answer to be inexact.

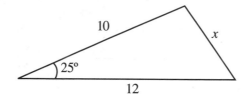

What is the value of *x*?

Solve this using the Law of Cosines. Remember that the angle *C* is opposite side *c* in the equation.

$c^2 = a^2 + b^2 - 2ab\,(\cos(C))$	Use the Law of Cosines.
$x^2 = 10^2 + 12^2 - 2 \times 10 \times 12\,(\cos(25°))$ $x^2 = 100 + 144 - 2 \times 10 \times 12\,(\cos(25°))$ $x^2 = 100 + 144 - 20 \times 12\,(\cos(25°))$ $x^2 = 100 + 144 - 240\,(\cos(25°))$ $x^2 \approx 100 + 144 - 240\,(0.906)$ $x^2 \approx 100 + 144 - 217.44$ $x^2 \approx 26.56$ $x \approx 5.15$	Plug in the known values. Note that the variable *x* from the figure replaces *c* in the equation. Simplify and solve. Use your calculator.

The first decision you have to make is which one (or both) of these laws is useful to you. You're trying to solve for a side where you have the opposite angle, but you don't have angles to match up with either of the other two sides you know. The sides are not equivalent, so you can't assume that the angles are equivalent. Consequently, you may not have enough information to use the Law of Sines. The Law of Cosines, on the other hand, would let you solve for the missing side, *c*, knowing only the other sides and the opposite angle. Line up each piece of the formula to find that $LJ^2 = 9^2 + 6^2 - 2(9)(6)\cos 50°$. Before you start calculating this value, glance at your answer choices—they aren't asking you to solve completely, just to match up the filled-in formula. Take the square root of both sides to find $LJ = \sqrt{9^2 + 6^2 - 2(9)(6)\cos 50°}$, or (E). The correct answer is (E).

ANSWERS TO CHAPTER 4 EXERCISES

Example 1 Answers (Page 157)

	$x°$	$y°$
Sin	$\dfrac{4}{5}$	$\dfrac{3}{5}$
Cos	$\dfrac{3}{5}$	$\dfrac{4}{5}$
Tan	$\dfrac{4}{3}$	$\dfrac{3}{4}$

Example 2 Answers (Page 158)

Angle	Complement
20°	70°
43°	47°
58°	32°

Example 3 Answers (Page 159)

$\sin(20°) \approx 0.342$	$\cos(70°) \approx 0.342$
$\cos(40°) \approx 0.766$	$\sin(50°) \approx 0.766$
$\sin(x°) = \dfrac{2}{5}$	$\cos(90 - x°) = \dfrac{2}{5}$
$\cos(90 - x°) = \dfrac{8}{9}$	$\sin(x°) = \dfrac{8}{9}$
$\tan(x°) = \dfrac{3}{4}$	$\tan(90 - x°) = \dfrac{4}{3}$
$\tan(90 - x°) = \dfrac{2}{3}$	$\tan(x°) = \dfrac{3}{2}$

Example 4 Answers (Page 160)

$x \approx \mathbf{53.13°}$

To solve for $x°$, you can enter $\sin^{-1}\left(\dfrac{4}{5}\right)$, $\cos^{-1}\left(\dfrac{3}{5}\right)$, or $\tan^{-1}\left(\dfrac{4}{3}\right)$.

$y \approx \mathbf{36.87°}$

To solve for $y°$, you can enter $\sin^{-1}\left(\dfrac{3}{5}\right)$, $\cos^{-1}\left(\dfrac{4}{5}\right)$, or $\tan^{-1}\left(\dfrac{3}{4}\right)$.

CHAPTER 4 PRACTICE QUESTIONS

Directions: Complete the following problems as specified by each question. For extra practice after answering each question, try using an alternative method to solve the problem or check your work.

1. In triangle XYZ, $\dfrac{\overline{XY}}{\overline{XZ}} = \dfrac{13}{5}$. Determine the values of the 6 trigonometric functions sin, cos, tan, csc, sec, and cot.

2. A window washer is cleaning the windows of a building. The ladder is fully extended to 25 feet and is leaning against the building. The base of the ladder is 7 feet away from the wall of the building.

    a. What is the vertical height, on the building, that the ladder reaches?

    b. What angle is created with the ground?

3. Given $\csc \theta = \dfrac{4}{3}$, determine the values of the other five trigonometric functions.

    Rationalize the denominators, if needed.

4. Given the triangle with the labeled side lengths, show that the following are true:

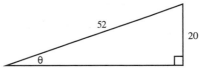

    a. $\sin \theta \times \sec \theta = \tan \theta$

    b. $\sin^2 \theta + \cos^2 \theta = 1$

5. Prove $\sin (90 - \theta) = \cos \theta$.

6. Prove $\tan (90 - \theta) = \dfrac{1}{\tan \theta}$.

7. For triangle ABC, $AB = 5$, $BC = 7$, and $m\angle A = 43°$. Find the exact value of $\angle C$; then approximate the value to 2 decimal places.

8. Triangle ABC has side lengths of $AB = 19$, $BC = 8$, and $AC = 14$. Use the lengths to find the three angles of the triangle.

SOLUTIONS TO CHAPTER 4 PRACTICE QUESTIONS

1. $\sin \theta = \dfrac{12}{13}$ $\cos \theta = \dfrac{5}{13}$ $\tan \theta = \dfrac{12}{5}$

   $\csc \theta = \dfrac{13}{12}$ $\sec \theta = \dfrac{13}{5}$ $\cot \theta = \dfrac{5}{12}$

The question provides the ratio for two sides of the right triangle, so use the Pythagorean theorem to determine the third side:

$a^2 + b^2 = c^2$	Pythagorean theorem
$(5)^2 + b^2 = (13)^2$	Substitute $a = 5$ and $c = 13$.
$25 + b^2 = 169$	Simplify.
$b^2 = 144$	Subtract 25 from both sides.
$b = 12$	Take the square root of both sides.

Note: Recall common Pythagorean triples to save time in calculations!

In terms of θ, the opposite side is 12, the adjacent side is 5, and the hypotenuse is 13. Determine the values of the 6 trigonometric functions:

$\sin \theta = \dfrac{opposite}{hypotenuse} = \dfrac{12}{13}$ $\csc \theta = \dfrac{hypotenuse}{opposite} = \dfrac{13}{12}$

$\cos \theta = \dfrac{adjacent}{hypotenuse} = \dfrac{5}{13}$ $\sec \theta = \dfrac{hypotenuse}{adjacent} = \dfrac{13}{5}$

$\tan \theta = \dfrac{opposite}{adjacent} = \dfrac{12}{5}$ $\cot \theta = \dfrac{adjacent}{opposite} = \dfrac{5}{12}$

2. It can be helpful to construct a figure that depicts the scenario taking place:

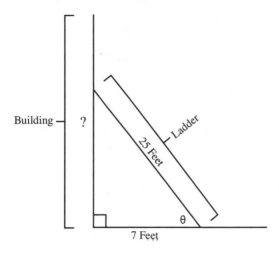

a. **24 feet**
   To determine the height, use the Pythagorean theorem:

$a^2 + b^2 = c^2$	Pythagorean theorem
$(7)^2 + b^2 = (25)^2$	Substitution using $a = 7$ and $c = 25$.
$49 + b^2 = 625$	Simplify.
$b^2 = 576$	Subtract 49 from both sides.
$b = 24$	Take the square root of both sides.

So, the height the ladder reaches is 24 feet.

b. $\approx$ **73.74°**
   To determine the angle the ladder creates with the ground, use the given lengths:

   Ladder distance from building = 7 feet (adjacent to θ)
   Ladder length   = 25 feet (hypotenuse)

   Use cosine to determine the angle:

$\cos \theta = \dfrac{adjacent}{hypotenuse}$	SOHCAHTOA
$\cos \theta = \dfrac{7}{25}$	Substitute given values
$\theta = \cos^{-1}\left(\dfrac{7}{25}\right)$	Take $\cos^{-1}$ of both sides
$\theta \approx 73.74°$	Use your calculator

3. $\sin \theta = \dfrac{3}{4}$

$\cos \theta = \dfrac{\sqrt{7}}{4}$

$\tan \theta = \dfrac{3\sqrt{7}}{7}$

$\csc \theta = \dfrac{4}{3}$ (given in problem)

$\sec \theta = \dfrac{4\sqrt{7}}{7}$

$\cot \theta = \dfrac{\sqrt{7}}{3}$

Recall that the cosecant is the reciprocal of the sine function, so $\sin \theta = 3/4$. One down, four more to go. Construct a right triangle to see what's going on:

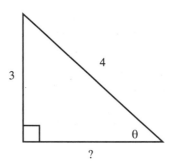

Since you know the opposite side, 3, and the hypotenuse, 4, use the Pythagorean theorem to determine the third side: (Note: This is not a 3-4-5 triangle! The hypotenuse is 4).

$a^2 + b^2 = c^2$	Pythagorean theorem
$(3)^2 + b^2 = (4)^2$	Substitution with $a = 3$ and $c = 4$.
$9 + b^2 = 16$	Perform arithmetic.
$b^2 = 7$	Subtract 9 from both sides.
$b = \sqrt{7}$	Take the square root of both sides.

So, now that the base of the triangle is known, use it to evaluate the remaining functions:

$\sin \theta = \dfrac{opposite}{hypotenuse} = \dfrac{3}{4}$

$\cos \theta = \dfrac{adjacent}{hypotenuse} = \dfrac{\sqrt{7}}{4}$

$\tan \theta = \dfrac{opposite}{adjacent} = \dfrac{3}{\sqrt{7}} \times \dfrac{\sqrt{7}}{\sqrt{7}} = \dfrac{3\sqrt{7}}{7}$

$\csc \theta = \dfrac{hypotenuse}{opposite} = \dfrac{4}{3}$ (given in problem)

$\sec \theta = \dfrac{hypotenuse}{adjacent} = \dfrac{4}{\sqrt{7}} \times \dfrac{\sqrt{7}}{\sqrt{7}} = \dfrac{4\sqrt{7}}{7}$

$\cot \theta = \dfrac{adjacent}{opposite} = \dfrac{\sqrt{7}}{3}$

4. Given the right triangle, use the two lengths to find the length of the third side:

$a^2 + b^2 = c^2$ — Pythagorean theorem
$(20)^2 + b^2 = (52)^2$ — Substitute $a = 20$ and $c = 52$.
$400 + b^2 = 2704$ — Simplify.
$b^2 = 2304$ — Subtract 400 from both sides.
$b = 48$ — Take the square root of both sides.

Note: 20:48:52 is the fourth multiple of 5:12:13.

Now, use the side lengths to prove the statements:

a. $\sin\theta \times \sec\theta = \tan\theta$ — Use SOHCAHTOA and the determined side lengths.

$$\left(\frac{20}{52}\right)\left(\frac{52}{48}\right) = \frac{20}{48}$$ — sin = opp/hyp, sec = hyp/adj, tan = opp/adj

$$\frac{20}{48} = \frac{20}{48}$$ — Simplify.

b. $\sin^2\theta + \cos^2\theta = 1$ — Use SOHCAHTOA and the determined side lengths.

$$\left(\frac{20}{52}\right)^2 + \left(\frac{48}{52}\right) = 1$$ — sin = opp/hyp, cos = adj/hyp

$$\frac{400}{2704} + \frac{2304}{2704} = 1$$ — Simplify.

$$\frac{2704}{2704} = 1$$

5. Construct an image to get an idea of what is going on:

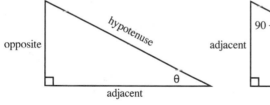 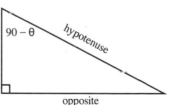

In a right triangle, the two acute angles always add up to 90°. So if one angle is θ, the other acute angle is always equal to 90 – θ.

The two triangles display the relationship of each side to the respective angle in question. Pay attention to the different labeling with the different angles in question. The triangles are identical, but the sides referred to as "opposite" and "adjacent" are different, depending on which angle you're referring to.

Now, label the angles in the same triangle and label the sides as *a*, *b*, and *c*:

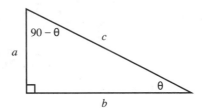

In this triangle, evaluate the required trig functions:

$\sin(90 - \theta) = \cos\theta$	Given
$\dfrac{opposite}{hypotenuse} = \dfrac{adjacent}{hypotenuse}$	SOHCAHTOA
$\dfrac{b}{c} = \dfrac{b}{c}$	Substitute the correct sides.

You could have also used a right triangle with numerical side lengths to prove this relationship.

6. Construct an image to get an idea of what is going on:

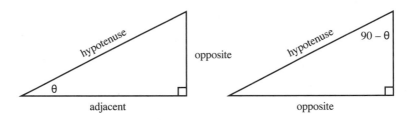

In a right triangle, the two acute angles always add up to 90°. So if one angle is θ, the other acute angle is always equal to 90 − θ.

The two triangles display the relationship of each side to the respective angle in question. Pay attention to the different labeling with the different angles in question. The triangles are identical, but the sides referred to as "opposite" and "adjacent" are different, depending on which angle you're referring to.

Now, label the angles in the same triangle and label the sides as a, b, and c:

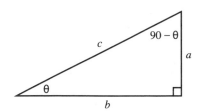

In this triangle, evaluate the required trig functions:

$$\tan(90 - \theta) = \frac{1}{\tan\theta}$$ Given

$$\frac{opposite}{adjacent} = \frac{1}{\frac{opposite}{adjacent}}$$ SOHCAHTOA

$$\frac{b}{a} = \frac{1}{\frac{a}{b}}$$ Substitute the correct sides.

$$\frac{b}{a} = \frac{b}{a}$$ Simplify.

You could have also used a right triangle with numerical side lengths to prove this relationship.

7. Given two side lengths and one angle of a triangle, use the Law of Sines:

$$\frac{\sin A}{a} = \frac{\sin B}{b} = \frac{\sin C}{c}$$ Law of Sines

Plug in the given values using a = 7, c = 5, and A = 43 to determine the exact value first:

$$\frac{\sin 43}{7} = \frac{\sin C}{5}$$ Substitute given values.

$$5 \times \frac{\sin 43}{7} = \sin C$$ Multiply both sides by 5.

$$\sin^{-1}\left(5 \times \frac{\sin 43}{7}\right) = C$$ Take the inverse sine of both sides.

This is the *exact* value as it does not have any rounding or approximation. Use a calculator (in degree mode!) to find the value of C and round to 2 decimal places:

$\sin^{-1}(5 \times \dfrac{\sin 43}{7}) = C$ Exact value

$29.15288537... = C$ Use your calculator.

$29.15 \approx C$ Approximate to 2 decimal places.

8. Given three side lengths of a triangle, use the Law of Cosines to determine angle measures. Since all three angles must be determined, use the side lengths interchangeably to determine the angles:

$a^2 = b^2 + c^2 - 2bc \cos A$
$b^2 = a^2 + c^2 - 2ac \cos B$ The Law of Cosines in all forms
$c^2 = a^2 + b^2 - 2ab \cos C$

Substitute the given values into each form to determine the angles:

$(8)^2 = (14)^2 + (19)^2 - 2(14)(19) \cos A$ Substitute the given values.

$64 = 196 + 361 - 532 \cos A$ Simplify.

$-493 = -532 \cos A$ Subtract $(196 + 361)$ from both sides.

$\dfrac{493}{532} = \cos A$ Divide both sides by -532.

$\cos^{-1}(\dfrac{493}{532}) = A$ Take the inverse cosine of both sides.

$22.08 \approx A$ Calculator approximation of angle measure.

Repeat the process to determine angle B:

$(14)^2 = (8)^2 + (19)^2 - 2(8)(19) \cos B$ Substitute the given values.

$196 = 64 + 361 - 304 \cos B$ Simplify.

$-229 = -304 \cos B$ Subtract $(64 + 361)$ from both sides.

$\dfrac{229}{304} = \cos B$ Divide both sides by -304.

$\cos^{-1}(\dfrac{229}{304}) = B$ Take the inverse cosine of both sides.

$41.12 \approx B$ Calculator approximation of angle measure.

Now, the Law of Cosines could be used one more time in the exact same process to yield the other angle, but remember that there are 180° in a triangle and two of the three angles are known. Subtract the known angle values to find the third!

$C + 22.08 + 41.12 = 180$ Sum of angles in a triangle equal 180°.

$C + 63.2 = 180$ Perform arithmetic.

$C = 116.8$ Subtract 63.2 from both sides.

So, the three angle measures are approximately 22.08°, 41.12°, and 116.8°.

REFLECT

**Congratulations on completing Chapter 4!
Here's what we just covered.
Rate your confidence in your ability to:**

- Understand and apply the basic trigonometric functions (sine, cosine, and tangent) and their reciprocals (secant, cosecant, and cotangent)

  ① ② ③ ④ ⑤

- Use trigonometric functions and their inverses on your calculator

  ① ② ③ ④ ⑤

- Find sine, cosine, and tangent values for complementary angles

  ① ② ③ ④ ⑤

- Use trigonometric functions to solve problems with right triangles

  ① ② ③ ④ ⑤

- Use the Pythagorean theorem to derive and understand additional trigonometric identities

  ① ② ③ ④ ⑤

- Use the Law of Sines and the Law of Cosines to solve problems with non-right triangles

  ① ② ③ ④ ⑤

If you rated any of these topics lower than you'd like, consider reviewing the corresponding lesson before moving on, especially if you found yourself unable to correctly answer one of the related end-of-chapter questions.

Access your online student tools for a handy, printable list of Key Points for this chapter. These can be helpful for retaining what you've learned as you continue to explore these topics.

Chapter 5
Three-Dimensional Figures

GOALS **By the end of this chapter, you will be able to:**

- Solve for area, circumference, radius, and diameter of a circle

- Find the incenter and circumcenter of a polygon

- Find the area or radius of a circle inscribed in a polygon, and vice versa

- Know the five types of regular polyhedra

- Understand cross-sections of prisms, spheres, cylinders, pyramids, and cones

- Know how to find volume and surface area of prisms, spheres, cylinders, pyramids, and cones

5

Lesson 5.1
From 2D to 3D

CIRCLES

In this lesson, we'll review important facts and formulas regarding circles. A **circle** is defined as the set of points located at a fixed distance from a given central point.

A point is "on" a circle if it is on the edge of the circle, not inside or outside of it. A circle has an infinite number of points.

The **radius** is the distance from the center to the edge of the circle. The term **radius** can also refer to a line segment that is drawn as such.

The **diameter** of a circle is the distance from edge to edge and through the circle's center. The term **diameter** can also refer to a line segment that is drawn as such. The diameter is twice as long as the radius.

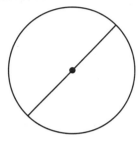

The **circumference** of a circle is the distance around the circle's edge. A circle's circumference is analogous to the **perimeter** of a polygon. The following formula gives the circumference of a circle:

Circumference of a Circle

Where r is the radius, and d is the diameter of a circle:

$$C = 2\pi r \quad \text{or} \quad C = \pi d$$

The value of π (a mathematical **constant** also known as **pi**) is actually defined by this relationship—the ratio of a circle's circumference to its diameter.

The **area** of a circle is given by the following formula:

> The value of π is approximately equal to 3.14. It's an irrational number, which means that it continues forever without repeating.

Area of a Circle

Where r is the radius of a circle:

$$A = \pi r^2$$

CARD

CARD is a mnemonic to remind you that all four of these elements of a circle—**C**ircumference, **A**rea, **R**adius, and **D**iameter—are related to one another. If you have just one of these values, you can easily solve for the other three.

Usually, it makes the most sense to solve for radius first; then plug radius into another formula to evaluate. The table below shows how to use the conventional equations to solve for radius.

Circumference $= 2\pi r$	Area $= \pi r^2$	Radius $= r$	Diameter $= 2r$
To solve for radius $C = (2\pi) \times r$ $\dfrac{C}{2\pi} = r$	To solve for radius $A = \pi \times r^2$ $\dfrac{A}{\pi} = r^2$ $\sqrt{\dfrac{A}{\pi}} = r$		To solve for radius $D = 2 \times r$ $\dfrac{D}{2} = r$

CARD Exercise

Try completing the table on the following page. For example, if area is given in a row, try solving for radius first; then solve for circumference and diameter. The first row has been completed. Answers are at the end of the chapter.

Circumference $= 2\pi r$	Area $= \pi r^2$	Radius $= r$	Diameter $= 2r$
12π	36π	6	12
			20
		7	
	25π		
6π			
		8	
	4π		
22π			
			8
		1	
	144π		
18π			

Here is how you may see circles on the ACT.

The circumference of a circle is 50 inches. How many inches long is its radius?

F. $\dfrac{25}{\pi}$

G. $\dfrac{50}{\pi}$

H. $\dfrac{100}{\pi}$

J. 50π

K. 100π

INCIRCLE AND CIRCUMCIRCLE

Incircle

The **incircle** of a polygon is a circle that is drawn inside the polygon and is **tangent** to every side. Another term for incircle is **inscribed circle.** The center of an incircle is called the **incenter**.

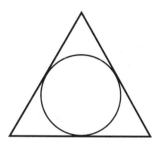

All triangles have incircles, as do all regular polygons. Irregular polygons with more than three sides will not necessarily have have an incircle that touches every side.

Some sources use a looser definition of "incircle," which can apply to all irregular polygons: "An incircle is the largest circle that can be drawn completely inside a given polygon."

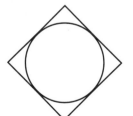 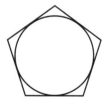 

Some of these polygons have an incircle, while others do not.

The radius of the incircle is known as the **apothem** or **inradius**. An apothem segment connects the incenter to a given side of the polygon, and is also perpendicular to that side. If a polygon has an incircle, then each side of the polygon has one apothem.

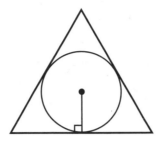

Circumcircle

The **circumcircle** of a polygon is a circle that is drawn outside the polygon and touches every vertex. Another term for circumcircle is **circumscribed circle**. The center of a circumcircle is called the **circumcenter**.

All triangles have circumcircles, as do all regular polygons. Irregular polygons with more than three sides will not necessarily have a circumcircle that touches every side.

The formula for the circumference of a circle is $C = 2\pi r$. So, $50 = 2\pi r$ and, therefore, $r = 25/\pi$. Choice (G) is the result of forgetting to divide by the 2 in the formula. Choice (H) could result by mistakenly multiplying 50 by 2 rather than dividing when using the formula. Choices (J) and (K) incorrectly solve for the radius. The correct answer is (F).

Some sources use a looser definition of "circumcircle," which can apply to all irregular polygons: "A circumcircle is the smallest circle that can be drawn completely outside a given polygon."

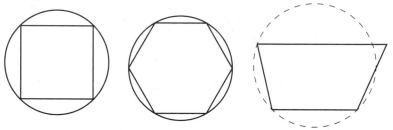

Some of these polygons have a circumcircle, while others do not.

The radius of the circumcircle is known as the **circumradius** or just **radius**. A circumradius segment connects the circumcenter to a given vertex of the polygon. If a polygon has a circumcircle, then each vertex of the polygon has a circumradius.

EXAMPLE

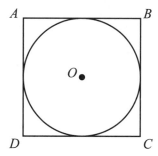

Which of the following statements is true? (Check all that apply.)

☐   The circle is inscribed in the square.

☐   The square is inscribed in the circle.

☐   The circle is circumscribed around the square.

☐   The square is circumscribed around the circle.

To answer this question, make the assumption that the circle is tangent to each side of the square. Incidentally, if that weren't true, then none of the above statements would be correct.

For this figure, the first and fourth statements are correct. The circle is inscribed in the square, and the square is circumscribed around the circle. Both statements mean the same thing.

WORKING WITH INSCRIBED FIGURES

Some geometry exercises will have you find the area (or other values) of inscribed or circumscribed figures. In general, when solving these problems, you should focus on places that the two figures intersect. In this lesson, we'll focus on a few examples of circles that are inscribed or circumscribed with polygons.

Circles and Squares

EXAMPLE

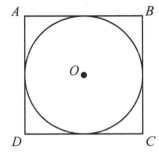

Supplies

Access your student tools to download larger, printable versions of the images in this section.

In the figure above, the circle with center *O* has a radius of 5 and is inscribed in square *ABCD*. Find the area of square *ABCD*.

First, draw the circle's radii at the points of intersection (the midpoints of the square's sides). By the definition of an inscribed circle, you know that the sides of the square are **tangent** to the circle, and each tangent line is perpendicular to the radius drawn at the point of tangency.

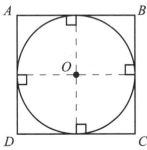

It's important to draw the radii at the points where the two figures intersect. If you draw radii anywhere else, it won't tell you much, because you'd have to figure out how to account for the gap between the circle and square.

Since *ABCD* is defined as a square, it's easy to see that the opposite midpoints of the square (also the tangent points for the circle) form **diameters** of the circle. That means that the diameter of the circle is congruent to the side of the square. (Recall that the radius of the circle was given as 5.)

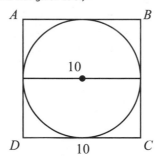

Therefore, the length of the side of the square is known. The area of a square is *side* × *side*, or *side*². For the area of this square, just multiply 10 × 10 = 100.

EXAMPLE 🔓

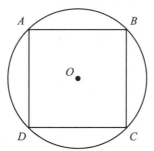

In the figure above, square *ABCD* has an area of 64 and is inscribed in the circle with center *O*. Find the area of circle *O*.

First, draw the circle's radii at the points of intersection (the square's vertices). This is a little different from the previous example, since the diameter of the circle is congruent to the square's *diagonals*, not its sides. You'll just have to do a couple more steps to solve for the length of the diagonal.

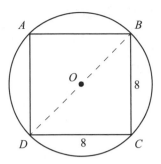

The square has a given area of 64. To solve for the side of the square, just calculate $\sqrt{64} = 8$. Then, use the Pythagorean theorem to solve for the diagonal:

$$a^2 + b^2 = c^2$$

$$8^2 + 8^2 = c^2$$

$$64 + 64 = c^2$$

$$128 = c^2$$

$$\sqrt{128} = c$$

$$\sqrt{64} \times \sqrt{2} = c$$

$$8\sqrt{2} = c$$

You may also recall the ratio for the sides of a 45°–45°–90° triangle. (This is also the triangle formed from the diagonal of a square).

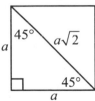

Therefore, the diagonal of this square is $8\sqrt{2}$, which is also congruent to the diameter of the circle. The radius of the circle would be equal to half of that, which is $4\sqrt{2}$. Now that you have the radius, you can solve for the area of the circle:

$$A = \pi r^2$$

$$A = \pi (4\sqrt{2})^2$$

$$A = \pi \times 4^2 \times (\sqrt{2})^2$$

$$A = \pi \times 16 \times (\sqrt{2})^2$$

$$A = \pi \times 16 \times 2$$

$$A = 32\pi$$

Circles and Equilateral Triangles

EXAMPLE **4**

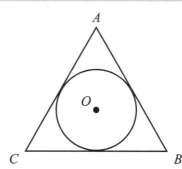

In the figure above, circle _O_ has a radius of 2 and is inscribed in the equilateral triangle _ABC_. Find the area of triangle _ABC_.

First, draw the circle's radii at the points of intersection (the midpoints of the triangle's sides). This forms three congruent quadrilaterals, with the angles 60°, 90°, 90°, and 120°. How do we know that? The 60° comes from the vertex angle of the triangle. Since the triangle is equilateral, we know that the vertex angles are each equal to 60°. The 90° angles are the ones formed at the tangent points. From the definition of tangent points, we know that the radii intersect at 90°. Finally, the 120° angle is what's left when you subtract 360° – (60° + 90° + 90°).

Here is how you may see inscribed figures on the ACT.

In the figure below, the circle with center _O_ is inscribed inside square _ABCD_ as shown. If a side of the square measures 8 units, what is the area of the shaded region?

F. 8 – 16π
G. 8π
H. 16π
J. 64 – 16π
K. 64π

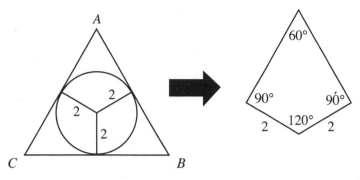

Recall that the sum of angles in a quadrilateral is 360°.

Next, if we divide these quadrilaterals in half, we get a special right triangle: the 30°–60°–90°. Recall that the ratio for the sides of such a triangle is $x : x\sqrt{3} : 2x$. We can use this to solve for the lengths of the sides of the small right triangle, which will give us information about triangle *ABC*.

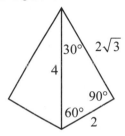

Perhaps the simplest way to find the area of triangle *ABC* is to find the area of this small right triangle, and multiply that by 6, since there are 6 such triangles that make up *ABC*. Find the area of the small right triangle.

$$A = \frac{1}{2}bh$$

$$A = \frac{1}{2} \times 2 \times 2\sqrt{3}$$

$$A = 1 \times 2\sqrt{3}$$

$$A = 2\sqrt{3}$$

Now, multiply that by 6 to find the area of *ABC*: $6 \times 2\sqrt{3} = 12\sqrt{3}$.

The area of triangle ABC is $12\sqrt{3}$.

Note that we can also find the lengths of the sides, as well as the height, of triangle *ABC*.

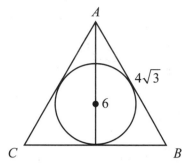

The length of each side is $4\sqrt{3}$.

Using these lengths to calculate the area of *ABC*, we get the same value of $12\sqrt{3}$.

$$A = \frac{1}{2}bh$$

$$A = \frac{1}{2} \times 6 \times 4\sqrt{3}$$

$$A = 3 \times 4\sqrt{3}$$

$$A = 12\sqrt{3}$$

The area of triangle ABC is $12\sqrt{3}$.

Mark the side of the square "8," and write down the formulas for the area of a circle and square: πr^2 and s^2.

Is there a formula for the shape made by the shaded region? Nope. We just need the basic formulas for the basic shapes. $8^2 = 64$, so we at least know the shaded region is less than 64, the area of the square. But what's the link between the square and the circle? The side of the square equals the diameter. So if the diameter is 8, then the radius must be 4. Use that in the area formula: $4^2\pi = 16\pi$. Subtract the area of the circle from the area of the square, and we get (J), the correct answer.

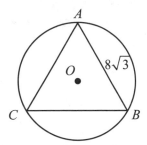

In the figure above, triangle *ABC* is equilateral with a side length of $8\sqrt{3}$, and it is inscribed in the circle with center *O*. Find the area of circle *O*.

First, draw the circle's radii at the points of intersection (the triangle's vertices). This forms three congruent triangles, with the angles 30°, 30°, and 120°. How do we know that? One way to think of it is that a circle has 360°, which we've divided into three equal parts at the center. We know that they are equal parts, because it is an equilateral triangle that is inscribed in the circle. Therefore, each of the three angles formed at the center of the circle is equal to 120° (= 360°/3).

We also know that each of these triangles is isosceles, since two of the sides are congruent to the radius of the circle. Therefore, the two smaller angles must be congruent. We can solve for this angle measure using an equation:

$$180° - 120° = 2x$$
$$60° = 2x$$
$$30° = x$$

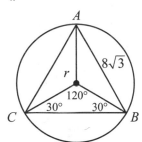

Next, if we divide these triangles in half, we get a special right triangle: the 30°–60°–90°. Recall that the ratio for the sides of such a triangle is $x : x\sqrt{3} : 2x$. We can use this to solve for the lengths of the sides of the small right triangle, which will give us information about circle *O*.

If the side length of triangle *ABC* is $8\sqrt{3}$, then half of that is $4\sqrt{3}$. For the small triangle, this is the length of the side opposite 60°.

From there, we can derive that the shortest side of the small triangle is 4, and the longest side is 8.

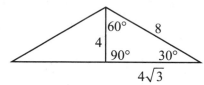

The long side of the small right triangle is congruent to the circle's radius. Therefore, the radius of circle *O* is 8. Calculate the area of the circle:

$$A = \pi\, r^2$$

$$A = \pi \times 8^2$$

$$A = \pi \times 64$$

$$A = 64\pi$$

The area of circle *O* is 64π.

Note that we can also find the height, as well as the area, of triangle *ABC*.

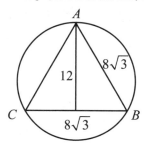

The height is 12. The area of *ABC* is ($\frac{1}{2} \times 12 \times 8\sqrt{3}$), or ($6 \times 8\sqrt{3}$).

Lesson 5.2
Three-Dimensional Figures

POLYHEDRA

A **polyhedron** (plural **polyhedra**) is a three-dimensional shape for which each face is a flat surface. In other words, each face is a **polygon**. The faces intersect each other as straight line segments, and these edges intersect each other at single-point vertices.

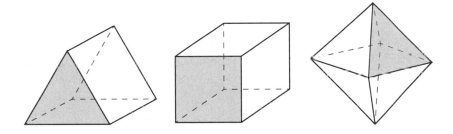

A **regular polyhedron** is one whose faces are all congruent, regular polygons. Also, in a regular polyhedron, the same number of faces meet at each vertex. There are only five types of regular, convex polyhedra, and they have special names (see the figures and descriptions on the next page).

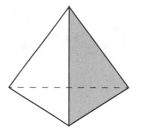

Tetrahedron

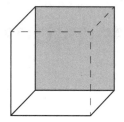

Hexahedron (Cube)

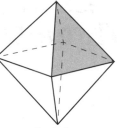

Octahedron

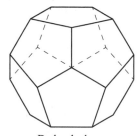

Dodecahedron

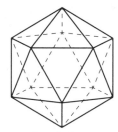

Icosahedron

- **Tetrahedron**—4 faces, which are equilateral triangles. Three triangles meet at each vertex.

- **Hexahedron (Cube)**—6 faces, which are squares. Three squares meet at each vertex.

- **Octahedron**—8 faces, which are equilateral triangles. Four triangles meet at each vertex.

- **Dodecahedron**—12 faces, which are regular pentagons. Three pentagons meet at each vertex.

- **Icosahedron**—20 faces, which are equilateral triangles. Five triangles meet at each vertex.

Why only five? Consider what happens when you connect two or more polygons at a single vertex. If you connect only two polygons, then you wouldn't have a solid. Therefore, to make a polyhedron, you need at least three polygons to meet at each vertex. Additionally, you need the sum of angles at each vertex to be less than 360°, because 360° is a flat plane, not a three-dimensional shape. This greatly limits the types of regular polyhedra that can be formed.

For example, if three squares meet at a single vertex, you may have part of a cube. If four squares meet at a single vertex, you would have a flat plane, not a solid. And, try as you might, you would never be able to connect five or more squares at a single vertex—it is simply impossible.

SPHERES

A **sphere** is a three-dimensional solid that is perfectly round. It is defined as the set of all points in space that are a given distance from its center. A sphere has no edges or vertices, and is not a polyhedron.

The distance from the center to the surface of a sphere is called the **radius**, and twice the radius is called the **diameter**.

A sphere is formed by rotating a circle (or semicircle) about its diameter.

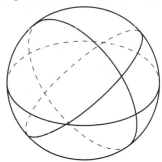

The **cross-section** of a sphere is always a circle. Cross-section is the term for the intersection of a plane through a solid, which forms a two-dimensional shape. Imagine slicing through an orange—the flat surface of the cut piece would be an approximate circle. If you slice through a perfectly spherical solid, the cross section would always be a perfect circle (not an **ellipse**), no matter what angle you cut it from.

A **great circle** is the largest circle that can be drawn around a given sphere. If you make a cross section that passes through the sphere's center, this forms a great circle. If a cross section does not pass through the sphere's center, that's known as a **small circle**.

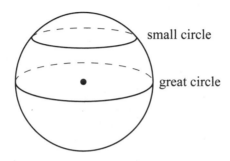

EXAMPLE 6

True or False? A great circle has the same radius as the sphere it inscribes.

True! If you find a great circle of a sphere, the radius of the great circle will be the same as the radius of the sphere.

Imagine cutting a sphere exactly in half, through its center. The cross-section of the cut face will be a circle with the same radius as the sphere. This is also the largest possible circular cross-section of the sphere.

PRISMS

A **prism** is polyhedron that has a pair of congruent, parallel faces (called the **bases** of the prism). The bases are on opposite ends of the prism, and the other faces (sometimes called **side faces**) are always parallelograms.

A prism is named for the shape of its base—for example, "triangular prism" or "pentagonal prism." In the figures below, the shaded parts show one base of each prism.

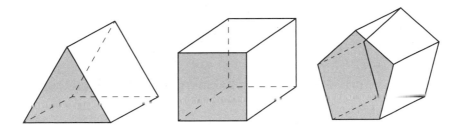

A prism is **regular** if its bases are regular polygons. Otherwise, the prism is **irregular**.

A prism is **right** if the bases are perpendicular to the other faces. Otherwise, it is an **oblique** prism. In a right prism, the side faces are always rectangles. All of the prisms in the above figure are right prisms.

Any cross-section parallel to the base will always be congruent to the base.

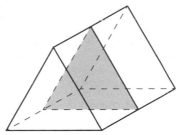

Other cross-sections will often form rectangles or parallelograms, but they may also form triangles, trapezoids, or other shapes. The number of sides in your cross-section is equal to the number of faces you slice through!

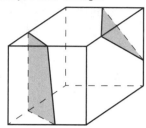

EXAMPLE

A cross-section of a cube is a polygon. How many sides could that polygon have? (Select all that apply.)

Reminder: A cube has six faces.

☐ 2

☐ 3

☐ 4

☐ 5

☐ 6

☐ 7

3, 4, 5, and 6 are correct. To get a three-sided cross section, you would intersect three faces of the cube. To get a four-sided cross section, you would intersect four faces of the cube. And so on. Remember, the number of sides in your cross-section is equal to the number of faces you slice through.

2 is not correct, because a polygon must have more than two sides! Moreover, if you "slice through" exactly two faces of a cube, you'd actually just be touching one of the **edges**, which gives you a line, not a polygon.

7 is not correct, because a cube has only six faces. For a seven-sided cross section, you would need a polyhedron with seven or more faces.

CYLINDERS

A **cylinder** is analogous to a prism, but its bases are circles. The **radius** of a cylinder is the radius of its circular base, and twice the radius is the **diameter**. The distance between the two bases is referred to as the **height**.

A cylinder is formed by rotating a rectangle about its edge, or about its center line.

Any cross-section parallel to the base will be a circle congruent to the base.

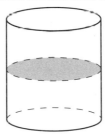

A cross-section perpendicular to the base will always be a rectangle.

Other, "slanted" cross-sections will be an **ellipse**, or a truncated ellipse (an ellipse with one or both ends cut off).

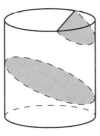

PYRAMIDS

A **pyramid** is a solid with a polygonal base, and triangular faces that meet at a vertex. This "top" vertex of the pyramid is known as the **apex**. The **height** or **altitude** of the pyramid is a line drawn from the apex perpendicular to the base. The **slant height**, conversely, is measured along a two-dimensional face.

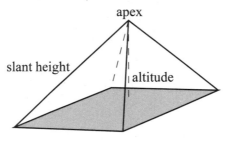

A pyramid is named for the shape of its base—for example, "triangular pyramid" or "pentagonal pyramid."

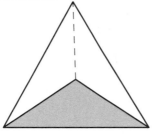 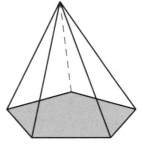

A pyramid is **regular** if its base is a regular polygon. Otherwise, the pyramid is **irregular**.

A pyramid is **right** if the apex is directly above the center of the base. Otherwise, it is an **oblique** pyramid.

In a right pyramid, any cross-section parallel to the base will be **similar** to the base, but smaller. The cross-section is **not** congruent to the base.

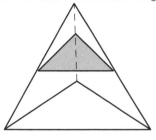 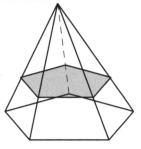

Other cross-sections may also form triangles, quadrilaterals, or other shapes.

CONES

A **cone** is analogous to a pyramid, but it has a circular base. Other than the base, a cone is considered to have one "side," which is curved. The top point of a cone is called the **apex**.

A cone is formed by rotating a right triangle about one of its legs. The **radius** of the base of the cone would be equal to the other leg of the triangle.

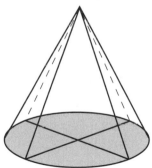

Any cross-section parallel to the base will be a circle, which is smaller than the base.

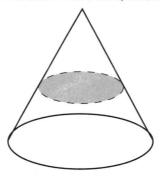

Other cross-sections will be ellipses or parabolas.

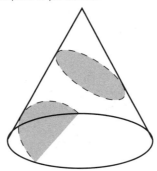

Lesson 5.3
Volume and Surface Area

Volume is the amount of 3-dimensional space that is occupied by a solid. You've probably noticed that beverages and other fluids are sold in containers marked for volume—quarts, gallons, and so on. You can also measure volume in "cubic" units, such as the cubic inch (a cube that is 1 inch on every side). In this chapter, you'll learn volume formulas for some of the most common types of 3-dimensional solids. You can use these formulas on more complex figures too, by breaking up a shape into smaller, more recognizable pieces.

Surface area is the amount of 2-dimensional area that is taken up by the **surface** of a figure. For example, the surface area of a cube is the sum of the areas of each of its six faces. In this chapter, you'll learn surface area formulas for some of the most common types of 3-dimensional solids. To calculate surface area on more complex figures, just remember that you need to calculate the area of every face on the figure.

PRISMS AND CYLINDERS

You can think of a prism as a two-dimensional shape that is stacked on top of itself to have a non-zero height. For example, a single sheet of sticky note paper can be thought of as a two-dimensional rectangle, but a whole pad of sticky notes would be a prism.

Volume

Some types of prisms have volume formulas that are fairly easy to remember. But whenever you don't know a formula for a certain prism, here is a general rule that works for every one: To calculate the volume for a prism, first find the area of the **base** (one of the two congruent faces on opposite sides of the figure); then multiply by the **height**.

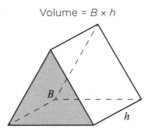

Volume of a Prism

The volume of any prism is equal to the **area** of the base, multiplied by the height of the prism.

Volume = $B \times h$

EXAMPLE 8

In the formula $V = Bh$, we use the capital letter for "B" to remind you that you're finding the *area* of the base, not just the length or the width.

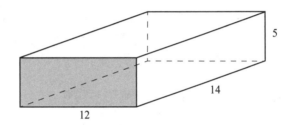

Find the volume of the rectangular prism shown above.

To calculate the volume of the prism, you'll first need to calculate the area of the base. Note that in a rectangular prism, you can actually consider any of the faces as the "base," since each pair of faces is parallel and congruent to each other.

Therefore, just pick any of the six faces and call it the base. For this example, we'll use the front face. This is a rectangle with width 5 and length 12. Calculate the area:

$$A = 5 \times 12$$

$$A = 60$$

Then, multiply that area by the height of the prism.

$$V = 60 \times 14$$

$$V = 840$$

To calculate the volume of a rectangular prism, you can use this straightforward formula:

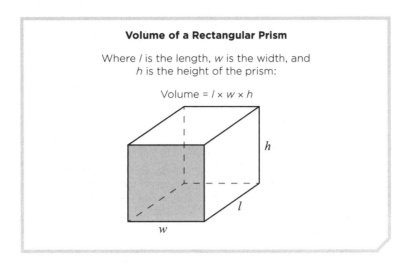

Volume of a Rectangular Prism

Where *l* is the length, *w* is the width, and *h* is the height of the prism:

Volume = $l \times w \times h$

Here is how you may see volume of a prism on the SAT.

If a rectangular swimming pool has a volume of 16,500 cubic feet, a uniform depth of 10 feet, and a length of 75 feet, what is the width of the pool, in feet?

A. 22

B. 26

C. 32

D. 110

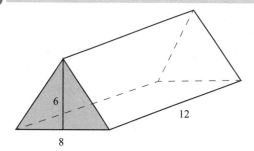

Find the volume of the triangular prism shown above.

To calculate the volume of the prism, you'll first need to calculate the area of the triangular base. Note that the triangles are considered the "bases" of this prism because they are congruent and parallel to each other. The other, rectangular faces are not parallel to each other, so they are not the bases.

The area of a triangle is given as $A = \frac{1}{2}bh$, and you know both the base and height of this triangle. Plug in the values and solve for A.

$$A = \frac{1}{2} \times 8 \times 6$$

$$A = \frac{1}{2} \times 48$$

$$A = 24$$

Finally, multiply that area by the height of the prism.

$$V = 24 \times 12$$

$$V = 288$$

From our calculations above, we can derive the following formula:

Volume of a Triangular Prism

Where *b* is the length of the triangular base, *a* is the altitude of the triangular base, and *h* is the height of the prism:

$$\text{Volume} = \frac{1}{2} \times a \times b \times h$$

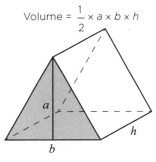

For this question, you need to know that volume equals *length* × *width* × *height*. You know that the volume is 16,500, the depth (or height) is 10, and the length is 75. Just put those numbers into the formula: 16,500 = 75 × *w* × 10. Use your calculator to solve for *w*, which equals 22: The correct answer is (A).

Volume of a Cylinder

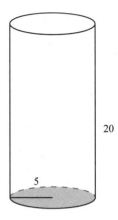

Find the volume of the cylinder shown above.

To calculate the volume of the cylinder, use the same basic idea as with a prism: Calculate the area of the base; then multiply by the height. First, calculate the area of the circular base.

The area of a circle is given as $A = \pi r^2$, and you know the radius for this circle.

$$A = \pi r^2$$

$$A = \pi \times 5^2$$

$$A = 25\pi$$

Finally, multiply that area by the height of the cylinder.

$$V = 25\pi \times 20$$

$$V = 500\pi$$

From our calculations above, we can derive the following formula:

Volume of a Cylinder

Where *r* is the radius of the circular base,
and *h* is the height of the cylinder:

$$\text{Volume} = \pi r^2 \times h$$

A cylindrical water tank has a diameter of 20 meters and a height of 20 meters. If the tank currently holds 1600 cubic meters of water, what is the approximate depth of the water in the tank?

A.   5 m

B.   10 m

C.   15 m

D.   20 m

Surface Area

As with volume, some types of prisms have formulas for surface area that are fairly straightforward. But whenever you don't know a formula for a certain prism, the general rule is to add up the areas of every face.

> **Surface Area of a Prism**
>
> The surface area (*SA*) of any prism is the sum of the areas of all of its faces.

EXAMPLE

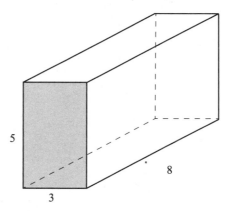

Find the surface area of the rectangular prism shown above.

To calculate the surface area of the prism, calculate the areas of its faces. Note that in a rectangular prism, each pair of opposite faces is congruent, so you'll only need to calculate the areas of three faces (or less, in some cases). The same would be true for an oblique rectangular prism.

Start with any face, and calculate the area. For this example, we'll start with the front face:

$$A = 3 \times 5$$
$$A = 15$$

Therefore, the area of the front face is 15, so you also know that the area of the back face is 15, since opposite faces are congruent.

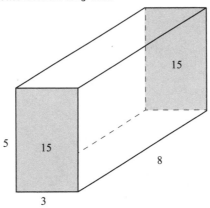

Find the area of a different face. We'll go with the one that's on the bottom:

$A = 3 \times 8$
$A = 24$

Therefore, the bottom and top faces each have an area of 24.

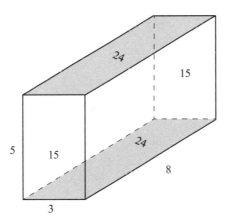

The height of the water tank doesn't matter in this problem. Use the formula for volume of a cylinder ($V = \pi r^2 \times h$) and plug in the information you have. It should look like this: $1600 = (\pi)(10^2)h$. When you solve for h, you get a number very close to 5. Since the question asks for the approximate depth of the water, (A) is the best answer.

The last pair of faces to calculate is the two side faces. Calculate the area of one of these sides:

$$A = 5 \times 8$$
$$A = 40$$

Therefore, the left and right faces each have an area of 40.

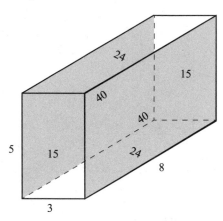

Finally, add all of these numbers together, making sure to include all six faces:

$$SA = 15 + 15 + 24 + 24 + 40 + 40$$

You can simplify first, if you like:

$$SA = 2 \times 15 + 2 \times 24 + 2 \times 40$$
$$SA = 2\,(15 + 24 + 40)$$
$$SA = 2 \times (79)$$
$$SA = 158$$

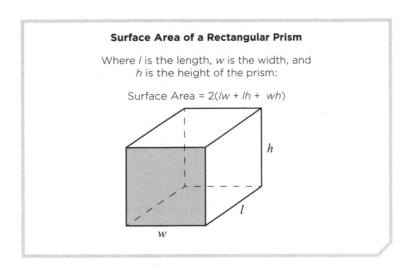

Surface Area of a Rectangular Prism

Where *l* is the length, *w* is the width, and *h* is the height of the prism:

Surface Area = $2(lw + lh + wh)$

EXAMPLE 12

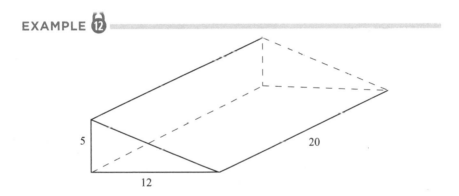

Find the surface area of the triangular prism shown above.

To calculate the surface area of the prism, we'll need to calculate the areas of its faces. Note that in a triangular prism, the pair of triangular bases is congruent, but the three rectangular faces may not be congruent.

The area of a triangle is given as $A = \frac{1}{2}bh$, but there are two of these triangles, so just simplify the pair to bh. Plug in the values and solve for area:

$A = bh$ For the **pair** of triangular bases

$A = 5 \times 12$

$A = 60$

Therefore, the two triangular bases together have an area of 60.

Next, find the area of each rectangular face. Looking at the side face, the length and width are 5 and 20. Calculate the area:

$A = 5 \times 20$

$A = 100$

The bottom face has length and width of 12 and 20. Calculate the area:

$A = 12 \times 20$

$A = 240$

The slanted face has a length of 20, but the width (the slant of the triangle) is not labeled. You'll have to calculate it, using the Pythagorean theorem.

$a^2 + b^2 = c^2$

$5^2 + 12^2 = c^2$

$25 + 144 = c^2$

$25 + 144 = c^2$

$169 = c^2$

$\sqrt{169} = c$

$13 = c$

Therefore, the width of the slanted side is 13. Calculate the area of the slanted triangular face:

$A = 13 \times 20$

$A = 260$

Finally, add the areas of all the faces together:

$SA = 60 + 100 + 240 + 260$

$SA = 660$

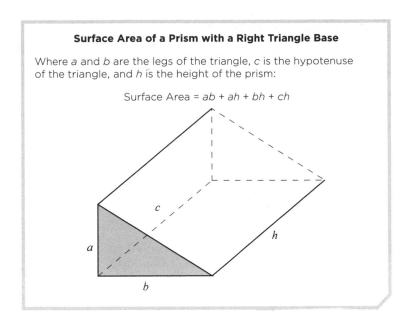

Surface Area of a Prism with a Right Triangle Base

Where a and b are the legs of the triangle, c is the hypotenuse of the triangle, and h is the height of the prism:

Surface Area = $ab + ah + bh + ch$

Prism Exercise

Can you come up with a similar formula for a prism whose base is a non-right triangle? Assume that the sides and altitude of the triangle are given, as well as the height of the prism.

Answers are at the end of the chapter.

Surface Area of a Cylinder

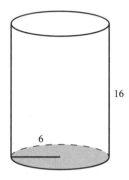

Find the surface area of the cylinder shown above.

To calculate the surface area of the cylinder, calculate the areas of its faces. But these faces are round! Here's how to do it.

Start with the circular bases, and calculate the area. Go ahead and multiply $\pi r^2 \times 2$, since there are two congruent circular faces.

$$A = \pi r^2 \times 2$$ For the **pair** of circular bases
$$A = \pi \times 6^2 \times 2$$
$$A = \pi \times 36 \times 2$$
$$A = \pi \times 72$$

Therefore, the area of the two circular bases together is 72π.

Now, calculate the area of the curved face. The height is given as 16, but what about the length? Imagine having a cardboard tube that you slice open. If you flatten out the tube, the length of the previously curved edge is equal to the **circumference** of the circle.

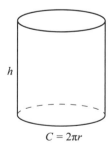

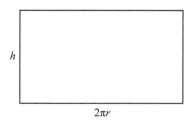

The circumference of a circle is given as $2\pi r$. Calculate the circumference of our circle:

$$C = 2\pi r$$
$$C = 2\pi \times 6$$
$$C = 12\pi$$

To calculate the area of the curved "face," multiply the circumference by the height of the cylinder.

$$A = 12\pi \times 16$$
$$A = 192\pi$$

Finally, add the areas together:

$$SA = 192\pi + 72\pi$$
$$SA = 264\pi$$

From our calculations above, we can derive the following formula:

Surface Area of a Cylinder

Where r is the radius and h is the height of the cylinder:

Surface Area = $2\pi (r^2 + rh)$

PYRAMIDS AND CONES

Volume

Once you are comfortable with finding the volume of a prism, finding the volume of a pyramid is very simple: Calculate the volume as though the figure is a prism; then divide by 3.

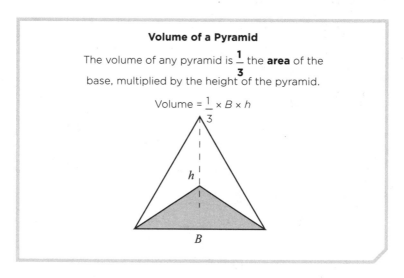

Volume of a Pyramid

The volume of any pyramid is $\frac{1}{3}$ the **area** of the base, multiplied by the height of the pyramid.

$$\text{Volume} = \frac{1}{3} \times B \times h$$

We can illustrate this relationship by combining three congruent pyramids to form a cube.

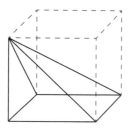 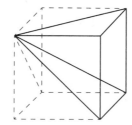 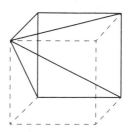

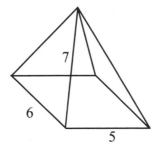

Find the volume of the rectangular pyramid shown above.

To calculate the volume of the pyramid, begin the same way you would if the figure were a prism—first, calculate the area of the base. Note that in a rectangular pyramid, the base is the rectangle, and all other faces are triangles. Calculate the area of the rectangular base:

$$A = 5 \times 6$$
$$A = 30$$

Then, multiply that area by the height of the pyramid.

$$30 \times 7 = 210$$

Then, divide by 3:

$$V = \frac{210}{3}$$
$$V = 70$$

To calculate the volume of a rectangular pyramid, you can use this formula:

Volume of a Rectangular Pyramid

Where l is the length, w is the width, and h is the height of the pyramid:

$$\text{Volume} = \frac{1}{3} \times l \times w \times h$$

EXAMPLE 15

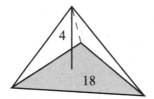

Find the volume of the triangular pyramid shown above.

In this example, the area of the base is already given, which simplifies things. In some other exercises, you would just need to calculate the area of the base first. Note that in a triangular pyramid, any of the faces can be considered the "base." However, it's important to remember that the height is perpendicular to the base, so make sure to calculate the height appropriately.

The area of the base is given:

$$A = 18$$

Multiply that area by the height of the pyramid:

$$18 \times 4 = 72$$

Then, divide by 3:

$$V = \frac{72}{3}$$
$$V = 24$$

Volume of a Cone

The process for calculating the volume of a cone is similar to that for a pyramid:
Calculate the volume as though the figure is a cylinder; then divide by 3.

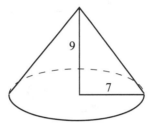

Find the volume of the cylinder shown above.

To calculate the volume of the cone, use the same basic idea as with a prism:
Calculate the area of the base, multiply that area by the height, and then divide by 3.
First, calculate the area of the circular base:

$$A = \pi r^2$$
$$A = \pi \times 7^2$$
$$A = 49\pi$$

Multiply that area by the height of the cone:

$$= 49 \times \pi \times 9$$
$$= 441\pi$$

Finally, multiply by $\dfrac{1}{3}$:

$$V = 441 \times \pi \times \frac{1}{3}$$
$$V = 147\pi$$

From our calculations above, we can derive the following formula:

Volume of a Cone

Where r is the radius of the circular base,
and h is the height of the cone:

$$\text{Volume} = \frac{1}{3} \times \pi r^2 \times h$$

Surface Area

To calculate the surface area of a pyramid, the most straightforward thing to do is to calculate the area of every face.

Surface Area of a Pyramid

The surface area of any pyramid is the sum
of the areas of all of its faces.

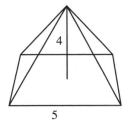

Find the surface area of the square, right pyramid shown above.

To calculate the surface area of the pyramid, calculate the areas of its faces. Start by calculating the area of the base:

$$A = 5 \times 5$$
$$A = 25$$

Therefore, the area of the bottom face is 25.

Note that in a **right regular** pyramid, all of the triangular faces are congruent. Therefore, we just need to calculate one of them.

$$A = \frac{1}{2} \times 5 \times 4$$
$$A = 10$$

Therefore, the triangular faces each have an area of 10.

Finally, add the areas of all the faces together:

$$SA = 25 + 10 + 10 + 10 + 10$$

$$SA = 65$$

Here is how you may see volume of a cone on the ACT.

The volume of a right circular cone with the bottom removed to create a flat base can be calculated from the following equation: $V = \frac{1}{3}\pi h(R^2 + r^2 + Rr)$ where h represents the height of the shape and R and r represent its radii as shown in the figure below:

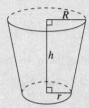

This formula can be used to determine the capacity of a large coffee mug. Approximately how many cubic inches of liquid can the cup shown below hold if it is filled to the brim and its handle holds no liquid?

A. 19
B. 50
C. 105
D. 109
E. 438

Surface Area of a Cone

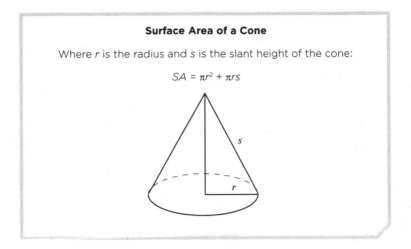

Surface Area of a Cone

Where *r* is the radius and *s* is the slant height of the cone:

$$SA = \pi r^2 + \pi r s$$

The formula for surface area of a cone is derived from the relationship between two circles. If you cut a cone open and flatten it out, you would have a portion of a circle:

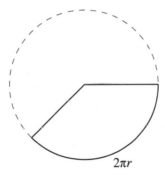

$2\pi r$

If we know the radius and arc length of a portion of a circle, then we can calculate the area of the "sector" of the circle. You will learn more about arcs and sectors in Chapter 7.

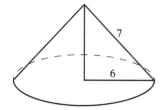

Find the surface area of the cone shown above.

Use the formula for the surface area of a cone:

$$SA = \pi r^2 + \pi r s$$
$$SA = \pi \times 6^2 + \pi \times 6 \times 7$$
$$SA = \pi \times 36 + \pi \times 42$$
$$SA = \pi \times (36 + 42)$$
$$SA = \pi \times 78$$

The second figure shows that the small and large diameters of the coffee cup are 4 and 6, respectively, so the radii are 2 and 3. Plugging the numbers into the equation given, $V = \frac{1}{3}\pi 5.5(3^2 + 2^2 + 3 \cdot 2) = \frac{1}{3}\pi 5.5(19) \approx 109$. Choice (A) is the portion of the equation inside the parentheses, and (C) neglects to multiply by $\frac{1}{3}\pi$. Choice (B) uses a height of $\frac{5}{2}$ instead of $5\frac{1}{2}$, or 5.5. Choice (E) plugs the diameters into the equation instead of the radii. The correct answer is (D).

POLYHEDRA

Since polyhedra can be all different shapes and sizes, there is no simple formula to apply to every one. Volume for a polyhedron is usually calculated by breaking up the figure into pyramids, and calculating the volume of each pyramid.

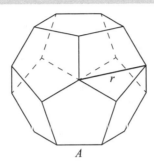

The figure above is a regular dodecahedron, with radius *r* and each face having area *A*. Find the volume area of the dodecahedron shown above.

Since regular polyhedra rarely have nice round numbers for both the side lengths and radius, this example is provided using variables. Here's how to work it.

First, consider each face as having a pyramid apex at the center of the figure. Such a pyramid would look like this:

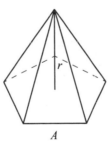

To calculate the volume of this pyramid, we would use the formula $V = \frac{1}{3}Bh$.

Substitute the variables from the figure to calculate the volume of each pyramid:

$$V = \frac{1}{3}Ar$$

Since this is a regular polyhedron, you know that each face would form such a pyramid that is congruent to the one above. Therefore, the volume of this dodecahedron is

$$V = \frac{1}{3}Ar \times 12$$
$$V = 4\ Ar$$

SPHERES

Volume of a Sphere

Where r is the radius:

$$V = \frac{4}{3}\pi r^3$$

EXAMPLE 20

Find the volume of a sphere with radius 6.

Use the formula for the volume of a sphere:

$$V = \frac{4}{3}\pi r^3$$
$$V = \frac{4}{3}\pi \times 6^3$$
$$V = \frac{4}{3}\pi \times 216$$
$$V = 288\pi$$

> **Surface Area of a Sphere**
>
> Where *r* is the radius:
>
> $$A = 4\pi r^2$$

The Greek philosopher Archimedes discovered this formula over two thousand years ago, when he theorized that the surface area of a sphere is equal to that of its inscribed cylinder.

EXAMPLE 21

Find the surface area of a sphere with radius 8.

Use the formula for the surface area of a sphere:

$$A = 4\pi r^2$$

$$A = 4 \times \pi \times 64$$

$$A = 256\pi$$

To see how spheres are tested on the SAT, access your Student Tools online.

ANSWERS TO CHAPTER 5 EXERCISES

CARD Exercise (Page 193)

Circumference = $2\pi r$	Area = πr^2	Radius = r	Diameter = $2r$
12π	36π	6	12
20π	100π	10	20
14π	49π	7	14
10π	25π	5	10
6π	9π	3	6
16π	64π	8	16
4π	4π	2	4
22π	121π	11	22
8π	16π	4	8
2π	1π	1	2
24π	144π	12	24
18π	81π	9	18

Prism Exercise (Page 227)

Where a is the altitude of the triangle, b is the base of the triangle, c and d are the other two sides of the triangle, and h is the height of the prism:

$$SA = ab + bh + ch + dh$$

CHAPTER 5 PRACTICE QUESTIONS

Directions: Complete the following problems as specified by each question. For extra practice after answering each question, try using an alternative method to solve the problem or check your work.

1. A square is inscribed in a circle with an area of 64π. What is the area of the square?

2. In the figure below, triangle *ABC* is inscribed in the circle with center *O*. What is the area of the shaded region?

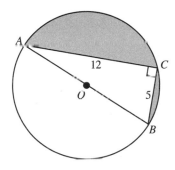

3. What is the radius of a cylinder with a volume of 108π and a height of 3?

4. What is the surface area of the cylinder described in question 3?

5. A pyramid has a volume of 40 cubic units and a base area of 20. What is the pyramid's height?

6. A hexagonal prism is composed of six identical equilateral triangular prisms. If the base of each triangular prism has an area of 12 cm² and the height of the hexagonal prism is 20 cm, what is the volume of the hexagonal prism?

7. Marcus wants to make himself an ice cream cone. He plans to pack the ice cream into the cone, filling it completely, and then add a scoop on top. If the cone has radius of 2 inches and a height of 6 inches and his ice cream scoop can make a spherical scoop with a radius of 3 inches, how many cubic inches of ice cream total does Marcus need?

8. Maria has constructed a square pyramid for a school project. She plans to cover the sides and base of the pyramid with contact paper. If the pyramid has a side length of 8 inches on the base and a slant height of 10 inches, how many square inches of contact paper will Maria need?

SOLUTIONS TO CHAPTER 5 PRACTICE QUESTIONS

1. **128**
   First draw the figure:

   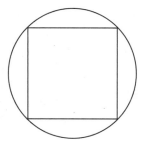

   Knowing the area of the circle makes it possible to find the radius. Use the formula $A = \pi r^2$ and fill in $64\pi = \pi r^2$. The π cancels out so $r^2 = 64$ and $r = 8$. No radius of the circle is entirely relevant to the square, but the diameter is. Draw a diameter that is a diagonal of the square.

   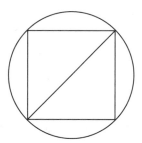

   That diameter has a length of 16 since the radius is 8. That is also a diagonal of a square. Since a square has equal sides, we can use the Pythagorean theorem: $x^2 + x^2 = 16^2$. Now $2x^2 = 256$, $x^2 = 128$, so $x = \sqrt{128}$, or $8\sqrt{2}$. That is the side of the square. The question asks for the area of the square, which is $\sqrt{128} \times \sqrt{128} = 128$. It is also possible to solve using the 45-45-90 triangle instead of the Pythagorean theorem.

2. **21.125π – 30 or about 36.366.**

   First, identify the shaded region. It is half of the circle minus the triangle. So we will need to find the area of the circle, divide that by 2, and then subtract the area of the triangle. Start by finding the area of the circle, for which the radius or diameter is needed. The diameter is the third side of the triangle. Use the Pythagorean theorem to find that the third side is 13 (or remember 5:12:13). That is the diameter, so the radius is half of 13, which is 6.5. Plug that into the area formula $A = \pi r^2$ to get $A = 42.25\pi$. That is the whole circle, so half of the circle has an area of 21.125π. Now find the area of the triangle. The base is 5 and the height is 12, so the formula is $A = 1/2(5)(12) = 30$. Thus, the area of the shaded region is $21.125\pi - 30$.

   One other strategy is to make a second right triangle to form one rectangle with length 12 and width 5. Find the area of the circle, subtract the area of the rectangle, and then divide by 2.

3. **6**

   Use the volume formula $V = B \times h$. The base is a circle, so its formula is πr^2. Thus, the full volume formula for a cylinder is $V = \pi r^2 h$. In this case, it is $108\pi = \pi r^2 3$. Cancel out the π: $108 = 3r^2$. Divide by 3: $36 = r^2$. Square root both sides to get $r = 6$.

4. **108π**

   The cylinder has a radius of 6 and height of 3. First, the surface area includes the top and bottom faces which are circles. Use the formula $A = \pi r^2$ to determine that each of the circles has an area of 36π for a total of 72π. The other part of the surface area is a rectangle with a length that is the circumference of the base and a height that is the height of the cylinder. Using the formula $C = 2\pi r$, the base has a circumference of 12π. Thus, the area of the rectangle is $A = l \times w = 12\pi \times 3 = 36\pi$. The total surface area is $72\pi + 36\pi = 108\pi$. You can also use the formula Surface Area $= 2\pi(r^2 + rh)$.

5. **6**

   Use the volume formula for a pyramid, which is $V = 1/3 \times B \times h$. Substitute the information provided: $40 = 1/3 \times 20 \times h$. Then simplify: $40 = 20/3h$. Multiply both sides by 3 to get $120 = 20h$. Finally, divide by 20 to get $h = 6$.

6. **1440 cm³**

   Find the volume of each triangular prism using the formula $V = B \times h$. The area of the base is 12 and the height is 20, so the formula is $V = 12(20) = 240$. That is for each triangular prism, so multiply by 6 to find that the total volume is 1440 cm³.

7. **44π**

   Since the question asks about cubic inches, it's a volume problem. First, find the volume of the cone using the formula $V = 1/3\pi r^2 h$. The radius is 2 and the height is 6, so that is $V = 1/3\pi(2^2)(6) = 1/3\pi(4)(6) = 1/3\pi(24) = 8\pi$. Now find the volume of the sphere using the formula $V = 4/3\pi r^3$. The radius is 3, so the formula is $V = 4/3\pi(3^3) = 4/3\pi(27) = 36\pi$. The total volume is $8\pi + 36\pi = 44\pi$ cubic inches.

8. **224**

Since the question involves square inches and "covering" something, it is a surface area problem. Find the surface area of the pyramid. Start with the base, which is a square, so its formula is $A = s^2$. The side is 8, so the area of the base is 64. A square pyramid also has four equal triangular faces. Use the formula $A = 1/2bh$. The base is 8 and the height is 10, so the formula is $A = 1/2(8)(10) = 40$. That is just for one triangular face, so multiply by 4 to find the total area for the triangular faces: 160. Now add the 64 square inches from the bottom to get a total of 224 square inches.

REFLECT

**Congratulations on completing Chapter 5!
Here's what we just covered.
Rate your confidence in your ability to:**

- Solve for area, circumference, radius, and diameter of a circle

  ① ② ③ ④ ⑤

- Find the incenter and circumcenter of a polygon

  ① ② ③ ④ ⑤

- Find the area or radius of a circle inscribed in a polygon, and vice versa

  ① ② ③ ④ ⑤

- Know the five types of regular polyhedra

  ① ② ③ ④ ⑤

- Understand cross-sections of prisms, spheres, cylinders, pyramids, and cones

  ① ② ③ ④ ⑤

- Know how to find volume and surface area of prisms, spheres, cylinders, pyramids, and cones

  ① ② ③ ④ ⑤

If you rated any of these topics lower than you'd like, consider reviewing the corresponding lesson before moving on, especially if you found yourself unable to correctly answer one of the related end-of-chapter questions.

 Access your online student tools for a handy, printable list of Key Points for this chapter. These can be helpful for retaining what you've learned as you continue to explore these topics.

Chapter 6
Connecting Algebra and Geometry

GOALS By the end of this chapter, you will be able to:

- Understand the standard form for the equation of a line

- Find an equation from a line, and plot a line from its equation

- Find the intersection of two lines

- Calculate the distance or midpoint between two coordinates

- Divide a segment to a specified ratio (e.g., 2:3 or 1:5)

- Find the vertex, roots, and axis of symmetry of a parabola

- Recognize the standard form and vertex form for the equation of a parabola

- Find the focus and directrix of a parabola from its equation

- Find the equation of a parabola given its zeroes, or its focus and directrix

- Prove geometric facts and theorems using algebra

Lesson 6.1
Algebra and the Coordinate Plane

COORDINATE GEOMETRY BASICS

Equation of a Line

> **Equation of a Line**
>
> $$y = mx + b$$

An equation in this form is called a **linear equation**, because it is guaranteed to form a straight line when graphed. The following the variables make up this equation:

- x and y are the coordinates (x, y) of a point on the line.

- m is the slope of the line (rise/run).

- b is the y-intercept of the line (the point where $x = 0$).

In most cases, when dealing with the equation of a line, you'll see x and y as variables, and numerical values for m and b. Such an equation represents all the possible (x, y) coordinates that exist on a specific line. Consider this example:

$$y = 3x + 4$$

In this example, any (x, y) coordinate pair that exists on the line will satisfy the equation, and any coordinate pair that satisfies the equation will exist on the line.

The equation is a function—if you plug in a value for x, you'll get a corresponding value for y. For instance, if we plug in $x = 2$, then $y = 10$:

$$y = 3(2) + 4$$
$$y = 6 + 4$$
$$y = 10$$

Therefore, we know that the coordinate (2, 10) exists on the line $y = 3x + 4$.

In coordinate geometry, you will often need to solve for one of the variables in a linear equation. For example, you might solve for slope (m) in order to find out if two lines are parallel. Or, you might solve for the x- or y-coordinate so that you can plot a point on a line.

EXAMPLE

The coordinate (2, 5) lies on the line with the equation $y = 3x + b$. What is the value of b?

In this example, there is only one unknown variable, b. If you plug in $x = 2$ and $y = 5$ (from the given coordinate), you can solve:

$$y = 3x + b$$
$$5 = 3(2) + b$$
$$5 = 6 + b$$
$$5 - 6 = b$$
$$-1 = b$$

Find a Line from an Equation

To graph a line, you just need to plot two points, and then connect them. You can find points that lie on a line by choosing different values for x and then evaluating for y, or vice versa.

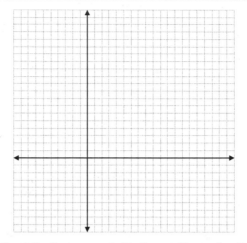

Supplies

If you're graphing with pencil and paper, it's usually a good idea to make your two points far apart if you can. This will help make your drawing more accurate.

Graph the line represented by the equation $y = 2x + 5$.

Begin by finding a coordinate pair that satisfies the equation. To find a coordinate pair, choose a value for x and then evaluate for y. Any value will do! Try to choose numbers that make the math easy.

If you plug in $x = 4$, then you get $y = 13$.

$$y = 2x + 5$$
$$y = 2(4) + 5$$
$$y = 8 + 5$$
$$y = 13$$

Therefore, the coordinate (4, 13) lies on the line.

Now, you just need one more coordinate pair to make a line. Again, any value will do. Try $x = -6$.

$$y = 2x + 5$$
$$y = 2(-6) + 5$$
$$y = -12 + 5$$
$$y = -7$$

Therefore, the coordinate (–6, –7) lies on the line.

Finally, plot the two points, and connect them to make a line.

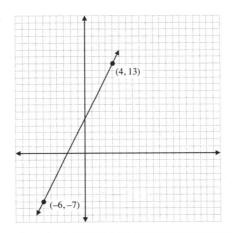

Find an Equation from a Line

To derive an equation from a line, you can start by choosing any two points from the line. With two points, you can then solve for *m* and *b*.

EXAMPLE 3

The coordinate (3, −4) and (−3, 10) both lie on the line with the equation
$y = mx + b$**. What is the value of** b**?**

First, plug in the two coordinate pairs to make two equations:

$$y = mx + b$$
$$-4 = m(3) + b$$
$$10 = m(-3) + b$$

Then, evaluate as a system of equations. One method is to add the two equations together and see if one of the variables cancels out.

$$-4 + 10 = m(3) + b + m(-3) + b$$
$$6 = m(3) + m(-3) + b + b \qquad \text{Combine like terms.}$$
$$6 = m(0) + 2b$$
$$6 = 2b \qquad\qquad\qquad m(0) \text{ cancels out.}$$
$$3 = b$$

Another method is to use the slope formula to solve for m first and then solve for b.

Slope Formula

To find the slope of a line containing the points (x_1, y_1) and (x_2, y_2):

$$\frac{y_1 - y_2}{x_1 - x_2}$$

Plug in the points $(3, -4)$ and $(-3, 10)$ to solve for slope:

$$\frac{y_1 - y_2}{x_1 - x_2}$$

Remember to keep your coordinates in the same order for the top and bottom of the fraction, or else you'll get the wrong sign.

$$= \frac{10 - (-4)}{(-3) - 3}$$

$$= \frac{14}{-6}$$

$$= -\frac{14}{6}$$

$$= -\frac{7}{3}$$

Therefore, $m = -\frac{7}{3}$. Plug that back into the equation:

$$y = mx + b$$

$$y = -\frac{7}{3}x + b$$

Finally, plug in one of the given coordinate pairs for x and y, and solve for b:

$$y = -\frac{7}{3}x + b$$

$$-4 = -\frac{7}{3} \times 3 + b$$

$$-4 = -7 + b$$

$$-4 + 7 = b$$

$$3 = b$$

Find the Intersection of Two Lines

If two lines intersect, they do so at a single point. To find that point, you can solve as a system of two equations.

EXAMPLE 4 🔒

Find the intersection of ($y = 4x - 3$) and ($y = 2x + 5$).

In Example 3, we solved a system of equations by adding the equations together. Another method for solving systems is called **substitution**—using an equivalent expression for a certain variable, so that you can solve for another variable.

If each of the equations has y alone, that makes it easy, because we can just set the expressions as equal to each other. We know that $y = y$, so the two expressions are also equal to each other.

$$y = 4x - 3$$
$$y = 2x + 5$$
$$4x - 3 = 2x + 5$$

Now the y variables are gone, and we are able to solve for x.

$4x - 3 = 2x + 5$	
$4x = 2x + 5 + 3$	Add 3 to both sides of the equation.
$4x = 2x + 8$	
$4x - 2x = 8$	Subtract $2x$ from both sides of the equation.
$2x = 8$	
$x = 4$	Divide both sides of the equation by 2.

We now have a value for x, but we still need y. Plug in $x = 4$ to either one of the original equations, and evaluate for y.

$$y = 4x - 3$$
$$y = 4 \times 4 - 3$$
$$y = 16 - 3$$
$$y = 13$$

Here is how you may see the equation of a line on the ACT.

The line with equation $5y - 4x = 20$ does NOT lie in which quadrant(s) of the standard (x, y) coordinate plane below?

A. Quadrant I only
B. Quadrant II only
C. Quadrant III only
D. Quadrant IV only
E. Quadrants I and III only

Quadrants of the standard (x, y) coordinate plane

Therefore, the intersection point is (4, 13).

Intersection points should always satisfy both equations, so you can check your work by plugging the coordinate pair into the other equation:

$$y = 2x + 5$$
$$13 = 2 \times 4 + 5$$
$$13 = 8 + 5$$
$$13 = 13 \qquad \checkmark \text{ True!}$$

Parallel Lines

EXAMPLE

In the standard coordinate plane, line *l* is represented by the equation $y + 9 = 7x$, and line *m* is represented by the equation $3y = 21x + 6$. Find the intersection, if any, of lines *l* and *m*.

Whenever you have a linear equation that's not already in the form of $y = mx + b$, a good first step is to put the equation in that form. This makes the equations much easier to compare.

For the first equation, *y* has a term added to it. To get *y* by itself, subtract 9 from both sides of the equation:

$$y + 9 = 7x$$
$$y = 7x - 9 \qquad \text{Subtract 9 from both sides of the equation.}$$

For the second equation, *y* has a coefficient. To get *y* by itself, divide both sides of the equation by 3:

$$3y = 21x + 6$$
$$y = \frac{21x + 6}{3} \qquad \text{Divide both sides of the equation by 3.}$$
$$y = 7x + 2 \qquad \text{Simplify (distributive property).}$$

Now, try setting the expressions equal to each other:

$$7x - 9 = 7x + 2$$
$$-9 = 2 \qquad \text{Subtract } 7x \text{ from both equations.}$$

If you subtract $7x$ from both equations, you get $-9 = 2$, which is obviously not a true equation. In other words, there is no value for *y* that can equal *both* quantities $(7x - 9)$ and $(7x + 2)$. This means that there is no solution for this pair of equations—no (x, y) coordinate pair that satisfies both equations.

If a pair of linear equations has no solution, that means that the two lines do not intersect at any point. And we know that when two lines do not intersect, it means that the lines are **parallel**. This is an algebraic explanation for the slopes of parallel lines being equal. For example, $(2x + 4)$ can't possibly have a value equal to $(2x + 9)$; therefore, we know that the pair of equations $(y = 2x + 4)$ and $(y = 2x + 9)$ has no solutions.

Slopes of Parallel Lines

If two lines are parallel, then they have the same slope, but different y-intercepts.

If two lines have the same slope, but different y-intercepts, then they are parallel.

Sometimes, you'll find that two equations are actually the same.

EXAMPLE 6

Determine the solutions, if any, for the lines $(2y = 6x - 7)$ and $(8y + 28 = 24x)$.

First, put both equations into the form of $y = mx + b$:

$$2y = 6x - 7$$

$$y = \frac{6x - 7}{2}$$ Divide both sides of the equation by 2.

$$y = 3x - \frac{7}{2}$$ Simplify (distributive property).

First, manipulate the equation given to get $y = \frac{4}{5}x + 4$. Then graph this line, with a slope of $\frac{4}{5}$ and a y-intercept of 4. The graph lies in the first, second, and third quadrants, and not in the fourth quadrant, (D)—the correct answer.

$$8y + 28 = 24x$$

$8y = 24x - 28$ Subtract 28 from both sides of the equation.

$y = \dfrac{24x - 28}{8}$ Divide both sides of the equation by 8.

$y = 3x - \dfrac{7}{2}$ Simplify (distributive property).

Therefore, these two equations actually represent the same line. If you graph both equations, they overlap each other entirely. The two equations have an **infinite** number of solutions.

Here is how you may see intersections of lines on the SAT.

Line m contains the points (4, 16) and (0, 8). At what point will line m intersect with line n if the equation of line n is $-8x + 4y = 24$?

    A. (0, 0)

    B. (−4, 0)

    C. These lines do not intersect.

    D. These lines intersect at an infinite number of points.

Lesson 6.2
Distance and Midpoint

In this lesson, we'll review how to calculate the distance and midpoint from two points in the coordinate plane. You'll also learn how to divide a line to a certain ratio, such as 1:3.

DISTANCE

The **distance** between two coordinates is the measure of the length between them. In the coordinate plane, you'll determine distance by using calculations, rather than a measuring tool such as a ruler.

One straightforward way to calculate distance is to use the Pythagorean theorem. In other words, you can construct a right triangle in which the two points form the hypotenuse. Then, use the Pythagorean theorem to find the length of the hypotenuse.

There are two ways that you can draw such a right triangle from two points—think of it like two halves of a rectangle. It does not matter which way you draw the triangle, because both are rotations of each other and are therefore congruent. The hypotenuse will be the same either way.

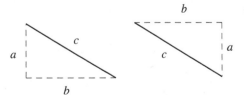

EXAMPLE 7

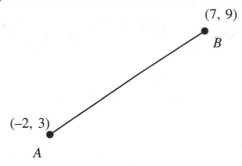

(7, 9)

B

(−2, 3)

A

Find the length of AB.

First, make a right triangle from the figure. Extend a horizontal line from one of the two points, and a vertical line from the other, to where they intersect.

SAT

In order to solve this problem, first put line n into slope-intercept form by adding $8x$ to both sides and then dividing both sides by 4. This will result in $y = 2x + 6$. You now know that the slope of line n is 2 and the y-intercept is 6. Since you also know the y-intercept of line m as it is given in point (0, 8), you know these are not the same lines. Eliminate (D).

To determine whether or not (C) will be the answer, find the slope of line m by using the slope formula:

$$\frac{8-16}{0-4} \quad \text{slope} = 2$$

Since both lines m and n have a slope of positive 2, they are parallel and will never intersect. Thus, (C) is correct.

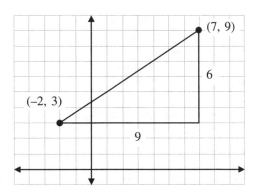

You can measure the legs of this triangle by counting, or by subtracting the x and y coordinates, respectively. The legs of this triangle are 6 and 9, as shown in the figure above.

Then, use the Pythagorean theorem to solve for the hypotenuse:

$$a^2 + b^2 = c^2$$

$$6^2 + 9^2 = c^2$$

$$36 + 81 = c^2$$

$$117 = c^2$$

$$\sqrt{117} = c$$

The length of AB is $\sqrt{117}$, or ≈ 10.82.

Distance Formula

The distance formula is derived from the Pythagorean theorem.

First, if we take the square root of both sides of the equation, we "solve" for c:

$$a^2 + b^2 = c^2$$

$$\sqrt{a^2 + b^2} = c$$

We can also substitute for a and b using the coordinates of the two points:

$$a = x_1 - x_2$$
$$b = y_1 - y_2$$

For the distance formula, you do not have to worry about the order of $(x_1 - x_2)$ versus $(x_2 - x_1)$—since you end up squaring these values, the result will always be positive.

Substitute those values for *a* and *b*, respectively:

$$c = \sqrt{a^2 + b^2}$$

$$c = \sqrt{\left(x_1 - x_2\right)^2 + \left(y_1 - y_2\right)^2}$$

That's the distance formula—or, you could call it "Pythagorean Theorem in the Coordinate Plane."

Distance Formula

To find the distance *d* between two points (x_1, y_1) and (x_2, y_2):

$$d = \sqrt{\left(x_1 - x_2\right)^2 + \left(y_1 - y_2\right)^2}$$

Here is how you may see coordinate distance on the ACT.

What is the perimeter of quadrilateral *STUR* if it has vertices with (x, y) coordinates $S\,(0, 0)$, $T\,(2, -4)$, $U\,(6, -6)$, $R\,(4, -2)$?

F. $2\sqrt{20}$
G. $2\sqrt{5} + 2\sqrt{20}$
H. $8\sqrt{5}$
J. 80
K. 400

Try plugging the points from the previous example, (2, 3) and (7, 9), into the distance formula:

$$d = \sqrt{\left(x_1 - x_2\right)^2 + \left(y_1 - y_2\right)^2}$$

$$d = \sqrt{\left(-2 - 7\right)^2 + \left(3 - 9\right)^2}$$

$$d = \sqrt{\left(-9\right)^2 + \left(-6\right)^2}$$

$$d = \sqrt{81 + 36}$$

$$d = \sqrt{117}$$

$$d \approx 10.82$$

This is the same result that we got when we used the Pythagorean theorem.

Memorizing the distance formula may help make your calculations easier, but either method is valid.

MIDPOINT

The **midpoint** of two coordinates is the point that lies exactly halfway between them.

Consider what you do when you need to find the "midpoint" of two *numbers*—you just take the average of the two numbers. For example, we know that 6 is exactly halfway between 2 and 10, because 6 is the average of 2 and 10: $\frac{2 + 10}{2} = 6$.

To find the midpoint between two coordinates, you're essentially doing the same thing—finding the average of the numbers. The difference is that there are two pairs of numbers instead of one. More specifically, you're finding the average of the *x*-coordinates and then the average of the *y*-coordinates.

Another way to think about midpoint is "the point that is halfway between the two *x*-values, and halfway between the two *y*-values."

EXAMPLE 8

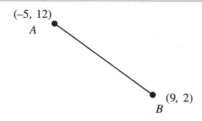

Find the midpoint of AB.

To calculate the midpoint between these coordinates, find the average of the x-values; then find the average of the y-values.

The average of the x-coordinates is $\dfrac{-5 + 9}{2}$, which equals 2. This is the x-coordinate of the midpoint.

The average of the y-coordinates is $\dfrac{12 + 2}{2}$, which equals 7. This is the y-coordinate of the midpoint.

Therefore, the midpoint is (2, 7).

ACT

Start by plotting out the figure to determine the distance between each point.

You can find the distance between each point by using the distance formula:

$d = \sqrt{\left(x_1 - x_2\right)^2 + \left(y_1 - y_2\right)^2}$. From point S to point R, for example, find

$d = \sqrt{(4 - 0)^2 + (-2 - 0)^2} = \sqrt{16 + 4} = \sqrt{20} = 2\sqrt{5}$. As you compare

the other sides, you'll notice that they all equal $2\sqrt{5}$, so the perimeter will be

$4 \times 2\sqrt{5} = 8\sqrt{5}$. The correct answer is (H).

Midpoint Formula

The Midpoint Formula is derived from taking the average of the x- and y-coordinates, respectively.

> **Midpoint Formula**
>
> To find the midpoint m between two points (x_1, y_1) and (x_2, y_2):
>
> $$\text{midpoint} = \left(\frac{x_1 + x_2}{2}, \frac{y_1 + y_2}{2} \right)$$

DIVIDING A SEGMENT TO A CERTAIN RATIO

Sometimes, instead of finding the midpoint, you'll need to divide a line segment to a different ratio, such as 1:3 or 2:5. For these exercises, we can perform the same procedure as with a dilation.

If you need to review dilations, see Chapter 3 of this book.

Here is how you may see midpoint on the ACT.

As shown below, the diagonals of rectangle *EFGH* intersect at the point (–2, –4) in the standard (x, y) coordinate plane. Point *F* is at (–7, –2). Which of the following are the coordinates of *H* ?

F. $\left(-4\frac{1}{2}, -3 \right)$

G. (–7, –6)

H. (3, –2)

J. (3, –6)

K. (–5, 3)

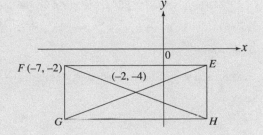

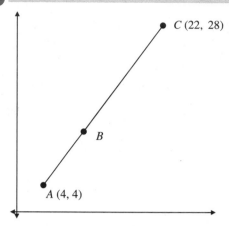

In the figure above, the ratio of *AB* to *BC* is 1:2. Find the coordinate for point *B*.

This example uses a ratio. Recall that ratios are a part-to-part relationship, while fractions are part-to-whole. A ratio of 1:2 is equivalent to a *fraction* of $\frac{1}{3}$. Another way to think about this ratio is with an algebraic expression for each length:

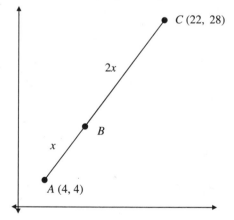

If *AB* has a length of *x*, then *BC* has a length of 2*x*. The total length of *AC* would be *x* + 2*x*, or 3*x*. That means that *x* is $\frac{1}{3}$ of the total length $\left(\frac{x}{3x}\right)$, while 2*x* is $\frac{2}{3}$ of the total length $\left(\frac{2x}{3x}\right)$.

Treat this exactly like a dilation, with point *A* as the center of dilation, and the scale factor as $\frac{1}{3}$.

Recall the formula for a dilation:

Dilation of a Coordinate

Where (x_1, y_1) is the center of dilation, and s is the scale factor, the dilated version of coordinate (x_2, y_2) is:

$$(x_1 + s(x_2 - x_1), y_1 + s(y_2 - y_1))$$

To perform the dilation of point C, plug in $(4, 4)$ as (x_1, y_1) and plug in $(22, 28)$ as (x_2, y_2). The scale factor is $\frac{1}{3}$.

$$(x_1 + s(x_2 - x_1), y_1 + s(y_2 - y_1))$$

$$\left(4 + \frac{1}{3}(22 - 4), 4 + \frac{1}{3}(28 - 4)\right)$$

$$\left(4 + \frac{1}{3}(18), 4 + \frac{1}{3}(24)\right)$$

$$(4 + 6, 4 + 8)$$

$$(10, 12)$$

Point B is at $(10, 12)$.

The point of intersection represents the midpoint between points F (–7, –2) and H (unknown). Note that the given midpoint is (–2, –4). You can use the midpoint formula, $\left(\dfrac{x_1 + x_2}{2}, \dfrac{y_1 + y_2}{2}\right)$ to determine the coordinates of point H. Solving for the x-coordinate gives $\left(\dfrac{-7 + x_2}{2} = -2\right)$. The x-coordinate of point H equals 3, eliminating (F), (G), and (K). Solving for the y-coordinate gives $\left(\dfrac{-2 + y_2}{2} = -4\right)$.

The y-coordinate of point H equals –6, eliminating (H). The correct answer is (J).

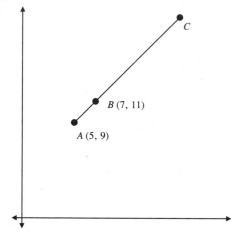

In the figure above, the ratio of *AB* to *BC* is 1:4. Find the coordinate for point *C*.

This example uses a ratio. To set this up as a dilation, first convert the ratio to a

fraction. Add the two parts to get the denominator: 1 + 4 = 5 parts. A ratio of 1:4 is

equivalent to a fraction of $\frac{1}{5}$. In this case, it means that *AC* is 5 times longer than *AB*.

Therefore, the scale factor is 5.

Treat this exactly like a dilation, with point *A* as the center of dilation, and the scale factor as 5.

Some ratio questions refer to "internal" or "external" division. Internal division means that the new point will be *between* the original two points. (Example 9 does this). External division means that the new point will form a longer version of the original line segment. (Example 10 does this).

To perform the dilation of point *B*, plug in (5, 9) as (x_1, y_1) and plug in (7, 11) as (x_2, y_2). The scale factor is 5.

$(x_1 + s(x_2 - x_1) , y_1 + s(y_2 - y_1))$

$(5 + 5 (7 - 5), 9 + 5 (11 - 9))$

$(5 + 5 (2), 9 + 5 (2))$

$(5 + 10, 9 + 10)$

$(15, 19)$

Point *C* is at (15, 19).

Lesson 6.3
Parabolas

A **parabola** is a U-shaped curve. It belongs to a group of objects called **conic sections**—literally, cross-sections of cones. A parabola is also what you get when you graph a **quadratic equation**—an equation in which one of the terms is raised to a power of 2 (for example, $y = x^2 + 2x + 4$).

The **vertex** is the point at which the parabola changes direction—that is, the curve swings from down to up, or left to right. All parabolas have a line of symmetry—called the **axis of symmetry**—that passes through the vertex.

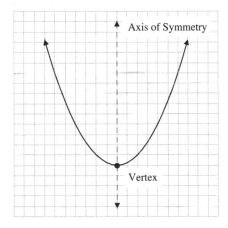

The definition of a parabola involves a point, called the **focus**, and a line, called the **directrix**. The parabola is the set of points that are equidistant from the focus and the directrix. In other words, if you have a point on a parabola (call the point X), the straight-line distance between X and the focus is equal to the perpendicular distance between X and the directrix.

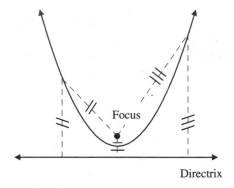

Note that the axis of symmetry of a parabola is perpendicular to the directrix, and passes through the focus and vertex. The vertex is always halfway between the focus and directrix.

GRAPHING A PARABOLA FROM AN EQUATION

Equation of a Parabola: Standard Form

$y = ax^2 + bx + c$ (vertical parabola)

$x = ay^2 + by + c$ (horizontal parabola)

You may have had some experience graphing parabolas in algebra class. One standard method is to first factor the quadratic equation, and solve for the **roots** (also known as **zeroes**) of the equation—that is, the x-values that make $y = 0$. This gives two points on the parabola, which you can then use to find the vertex. For the purposes of this lesson, we'll assume that you're somewhat familiar with factoring quadratic expressions. If not, you may find it helpful to brush up a little before proceeding.

Graph the parabola represented by the equation $y = x^2 + 8x + 15$.

Try factoring the quadratic expression. Begin by finding the factors of 15:

1, 15
3, 5

Now, find the pair of factors that adds up to 8.

1 + 15 = 8?	No.
3 + 5 = 8?	Yes!

That means that the binomial factors of $x^2 + 8x + 15$ are $(x + 3)$ and $(x + 5)$.

$y = x^2 + 8x + 15$
$y = (x + 3)(x + 5)$

Now, find the values for x where $y = 0$.

$0 = (x + 3)(x + 5)$

That means that either $(x + 3)$ or $(x + 5)$ equals zero.

If $(x + 3) = 0$, then $x = -3$.
If $(x + 5) = 0$, then $x = -5$.

Plugging in either of those two values into the original equation will make $y = 0$.

$y = (x + 3)(x + 5)$
$\quad = (-3 + 3)(-3 + 5)$
$\quad = (0)(2)$
$\quad = 0$

and

$y = (x + 3)(x + 5)$
$\quad = (-5 + 3)(-5 + 5)$
$\quad = (-2)(0)$
$\quad = 0$

Therefore, the two roots of the equation are $x = -3$ and $x = -5$. In other words, the coordinates (-3, 0) and (-5, 0) exist on the parabola, where the parabola intersects the x-axis.

The parabola's vertex will have an x-value that is exactly halfway between the two zeroes. (This is also true using any two points that have the same y-value on the parabola). The coordinate that's exactly between (-3, 0) and (-5, 0) is (-4, 0).

Therefore, the *x*-coordinate of the vertex is −4. To find the corresponding *y*-value, plug −4 into the equation.

$$y = (x + 3)(x + 5)$$
$$y = (-4 + 3)(-4 + 5)$$
$$y = (-1)(1)$$
$$y = -1$$

Therefore, when *x* = −4, *y* = −1. This gives us the coordinate (−4, −1), which is the parabola's vertex. With the vertex and two zeroes, we can graph the parabola:

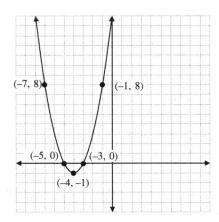

You may feel more comfortable with your graph if you plug in to find a few more points. For example, if *x* = −1, then *y* = 8 ((−1 + 3)(−1 + 5)); and if *x* = −7, then *y* = 8 ((−7 + 3)(−7 + 5)).

Here is how you may see parabolas on the SAT.

What is the vertex of the parabola defined by the equation $y = x^2 + 4x - 12$?

A. (0, −12)

B. (−6, 0)

C. (−2, −16)

D. (2, −12)

Equation of a Parabola: Vertex Form

For a parabola with vertex (h, k):

$y = a(x - h)^2 + k$ (vertical parabola)
$x = a(y - k)^2 + h$ (horizontal parabola)

An equation in **vertex form** provides specific information—namely, the coordinate of the vertex. If you see an equation in this form, you can know immediately that the vertex is (h, k).

However, the vertex by itself is not enough to tell us the shape of the parabola. To plot the parabola, you'll need at least three points. Therefore, choose a couple of different values for x and find the corresponding y-coordinates.

EXAMPLE

Graph the parabola with the equation $y = 2(x - 4)^2 + 3$.

Since this equation is in vertex form, you know that the vertex of the parabola is $(4, 3)$.

To graph the rest of the parabola, you'll need a couple more points. To find additional points that lie on the parabola, you can choose different values for x and then evaluate for the corresponding y-coordinates.

Let's try $x = 2$.

$$y = 2(x - 4)^2 + 3$$
$$y = 2(2 - 4)^2 + 3 \quad \text{Plug in } x = 2.$$
$$y = 2(-2)^2 + 3 \quad \text{Simplify.}$$
$$y = 2(4) + 3$$
$$y = 8 + 3$$
$$y = 11$$

Therefore, the point $(2, 11)$ lies on the parabola.

Try one additional value for x. We'll use $x = 6$.

$$y = 2(x - 4)^2 + 3$$
$$y = 2(6 - 4)^2 + 3 \qquad \text{Plug in } x = 6.$$
$$y = 2(2)^2 + 3 \qquad \text{Simplify.}$$
$$y = 2(4) + 3$$
$$y = 8 + 3$$
$$y = 11$$

Therefore, the point (6, 11) also lies on the parabola.

With three points, the parabola can be graphed:

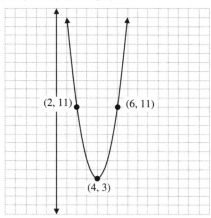

You can find the vertex by remembering that a parabola is symmetrical. If you find the x-values for a given y-value, the midpoint between the x-values will be the same as the x-value at the vertex. The easiest point to find is when $y = 0$. You can factor $x^2 + 4x - 12 = 0$ as $(x + 6)(x + 2) = 0$. Therefore, $x = -6$ or 2; the midpoint is the average of the x-values: $\dfrac{-6 + 2}{2} = -2$. Only (C) has an x-coordinate of -2. The correct answer is (C).

You can also derive the focus and directrix from the vertex form of an equation.

> **Focus and Directrix from Vertex Form**
>
> For a parabola with vertex (h, k):
>
> $y = a(x - h)^2 + k$ (vertical parabola)
>
> $x = a(y - k)^2 + h$ (horizontal parabola)
>
> The value $\dfrac{1}{4a}$ is equal to the distance between the vertex and the focus. This is also equal to the distance between the vertex and the directrix.

Therefore, if you have a parabola in vertex form, you can add the quantity $\dfrac{1}{4a}$ to the vertex coordinate in order to find the focus.

Likewise, you can subtract $\dfrac{1}{4a}$ from the vertex coordinate to find the y-coordinate of the directrix (for a vertical parabola).

EXAMPLE 13

Find the vertex and directrix for the parabola with the equation $y = \left(\dfrac{1}{8}\right)(x - 4)^2 + 3$.

This equation is already in vertex form, and you know that $a - \dfrac{1}{8}$. Find the quantity $\dfrac{1}{4a}$:

$$\frac{1}{4a} = \frac{1}{4\left(\dfrac{1}{8}\right)} = \frac{1}{\dfrac{4}{8}} = \frac{1}{\dfrac{1}{2}} = 2$$

Therefore, the distance between the vertex and the focus is 2. The vertex is $(4, 3)$, from the equation. Since this is a vertical parabola (it's in the form of $y =$), just add 2 to the y-coordinate.

Vertex = $(4, 3)$
Focus = $(4, 5)$

The directrix, then, will be two units below the vertex, so it passes through the coordinate $(4, 1)$. Remember that the directrix is a line, which in this case is the line $y = 1$.

Directrix: $y = 1$

FINDING THE EQUATION OF A PARABOLA

Find the Equation, Given the Zeroes

To find the equation of a parabola from its graph, try to start with the zeroes—that is, the points where the parabola crosses the x-axis. This will give you two factors of the equation. For the purposes of this exercise, let's say that we have two points on the parabola at $(m, 0)$ and $(n, 0)$. Then, we know that two factors are $(x - m)$ and $(x - n)$.

However, that's not enough to derive the full equation, because there are actually an infinite number of parabolas that can pass through those two points.

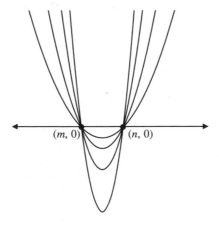

Therefore, we need a third point in order to find the missing factor, a. You're going to put your equation in the following form, and plug in the values of a known coordinate to solve for a.

$$y = a\,(x - m)(x - n)$$

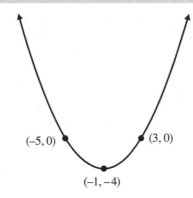

(−5, 0)   (3, 0)

(−1, −4)

Find the equation of the parabola with zeroes at (−5, 0) and (3, 0), and the vertex (−1, −4).

If the zeroes are (−5, 0) and (3, 0), then you know that two factors of the equation are $(x + 5)$ and $(x − 3)$.

$$y = a\,(x + 5)(x − 3)$$

To solve for a, plug in the coordinates of the vertex (that is, $x = −1$ and $y = −4$).

$y = a\,(x + 5)(x − 3)$

$−4 = a\,(−1 + 5)(−1 − 3)$ Plug in $x = −1$ and $y = −4$.

$−4 = a\,(4)(−4)$ Simplify.

$−4 = a\,(16)$

$\dfrac{−4}{16} = a$ Divide both sides of the equation by 16.

$\dfrac{1}{4} = a$ Simplify.

Therefore, the equation of the parabola is $y = \dfrac{1}{4}(x + 5)(x − 3)$.

Find the Equation, Given the Focus and Directrix

Recall that any point on a parabola is equidistant from the focus and the directrix. That is, if you have a point (x, y), then the distance to the focus is equal to the distance to the directrix. Therefore, to derive the equation of a parabola, we can just find those two distances and set them equal to each other.

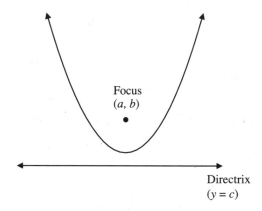

In the figure above, the focus of the parabola is (a, b) and the directrix is the line $y = c$. Take a point on the parabola, (x, y). To calculate the distance from (x, y) to (a, b), use the distance formula or Pythagorean theorem:

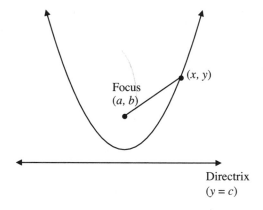

Distance from (x, y) to (a, b):

$$d = \sqrt{\left(x - a\right)^2 + \left(y - b\right)^2}$$

Now, find the distance between (x, y) and the directrix. Recall that the shortest distance between a point and a line will be through a segment perpendicular to the line, as shown:

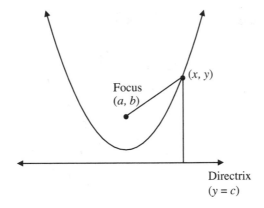

Which means that the distance between (x, y) and the directrix is $|y - c|$. (Use absolute value, because distances are always positive).

Now, set the values equal to each other, and simplify. Yes, this will have a lot of steps! Work carefully, and don't do anything in your head.

$$|y - c| = \sqrt{(x - a)^2 + (y - b)^2}$$

$(y - c)^2 = \left(\sqrt{(x - a)^2 + (y - b)^2} \right)^2$ Square both sides of the equation.

$(y - c)^2 = (x - a)^2 + (y - b)^2$ The root sign cancels out.

$y^2 - 2yc + c^2 = (x - a)^2 + (y - b)^2$ Expand $(y - c)^2$.

$y^2 - 2yc + c^2 = (x - a)^2 + y^2 - 2yb + b^2$ Expand $(y - b)^2$.

$y^2 + 2yb - 2yc + c^2 = (x - a)^2 + y^2 + b^2$ Add $2yb$ to both sides.

$y^2 + 2yb - 2yc = (x - a)^2 + y^2 + b^2 - c^2$ Subtract c^2 from both sides.

$2yb - 2yc = (x - a)^2 + b^2 - c^2$ Subtract y^2 from both sides.

$2y(b - c) = (x - a)^2 + b^2 - c^2$ Factor $2y$ from the left side of the equation.

$$y = \frac{(x - a)^2 + b^2 - c^2}{2(b - c)}$$ Divide both sides of the equation by $2(b - c)$.

That's quite a bit of algebra, there, and it can still be simplified further! But this format is fine, and it has y by itself.

Equation of a Parabola, with Focus and Directrix

Given focus (a, b) and directrix $y = c$:

$$y = \frac{(x - a)^2 + b^2 - c^2}{2(b - c)}$$

Lesson 6.4
Algebraic Proofs

In this lesson, you'll learn more about proofs and how to use algebra to complete proofs. When you are instructed to complete an **algebraic proof**, it means that you are expected to rely almost entirely on algebraic expressions and equations, and not a specific example with numerical values. The idea is usually that you're constructing a proof that works for *any* scenario, not just with a certain set of numbers. As a reminder, there are usually several different possible ways that you can set up a proof. There is no "right" or "wrong" approach, as long as you follow the instructions set by the task, and show all of your steps from beginning to end.

EXAMPLE 🔒 15

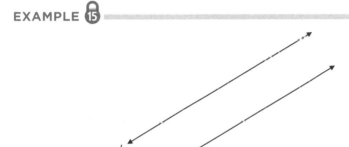

l_2

l_1

Your friend, a sculpture artist, is making a sketch of her next project which will use mirrors and lasers. She draws two parallel lines, which represent two large mirrors. She's certain that these lines are parallel, because she measured a constant distance between them. However, she's not sure if that means that the lines will have the same slope. She needs to be certain of this, in order to achieve the desired effect with the lasers. She asks you to explain it to her algebraically, so that she knows her sketch will work.

Can you prove algebraically that two parallel lines will always have the same slope?

One way to complete this proof is to use similar triangles. This works because we'll be able to use the legs of the triangles to represent the slope. Construct both a horizontal and vertical transversal intersecting the lines, as shown:

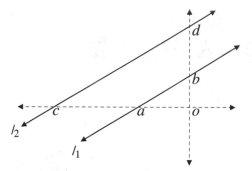

The transversals form triangles with our parallel lines, and we can prove that these triangles have corresponding, congruent angles. Using the horizontal transversal, we know that *cdo ≈ abo*. And from the vertical transversal, we know that $\angle dco \approx \angle bao$. Both of those pairs are alternate interior angles. Additionally, $\angle cod \approx \angle aob$, because these represent the same angle shared by both triangles.

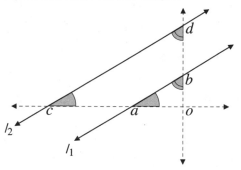

Since the two triangles have three pairs of corresponding, congruent angles, we know that the triangles are similar.

Now, consider what we know about the slopes of these lines. The two triangles are right triangles, since the horizontal and vertical transversals necessarily intersect each other at a right angle. Therefore, we can use the legs of the right triangles to represent the rise and run of the lines.

$$\text{Slope } l_1 = \frac{\text{rise}}{\text{run}} = \frac{\overline{bo}}{\overline{ao}}$$

$$\text{Slope } l_2 = \frac{\text{rise}}{\text{run}} = \frac{\overline{do}}{\overline{co}}$$

And using the definition of similar triangles, we know that $\frac{\overline{bo}}{\overline{ao}} = \frac{\overline{do}}{\overline{co}}$. Thus, it is proved that the two slopes are equal.

Given: $l_1 \perp l_2$

m_1 is the slope of l_1

m_2 is the slope of l_2

Prove: $m_1 \times m_2 = -1$

To review rotations, see Chapter 1 of this book.

One way to complete this proof is to use the rules for rotations of coordinates. This works because any point on l_1 would be a 90° rotation of a corresponding point on l_2, and vice versa.

First, we'll say that we've translated l_1 and l_2 so that their intersection is at the origin. We know that any time we translate a line, the slope will stay the same. Draw a figure with two perpendicular lines which intersect at the origin:

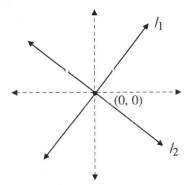

Choose a point on l_2 and call it (a, b). If we rotate this point 90° about the origin, the rotated coordinate will be $(-b, a)$.

This rotated coordinate is sure to lie on line l_2, because we know that l_1 and l_2 are perpendicular, and thus they meet at right angles.

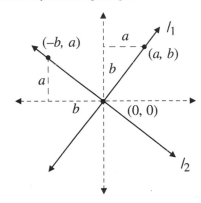

Now, since each of these lines has expressions for two coordinates (including the origin), we can calculate their slopes.

Slope of l_1:

$$= \frac{b - 0}{a - 0}$$

$$= \frac{b}{a}$$

Slope of l_2:

$$= \frac{a - 0}{-b - 0}$$

$$= \frac{a}{-b}$$

Now, find that $m_1 \times m_2 = -1$:

$$\frac{b}{a} \times \frac{a}{-b}$$

$$= \frac{ba}{-ab}$$

$$= -1 \qquad\qquad \text{Simplify.}$$

To set up an algebraic proof for a geometric figure, ask yourself what you know about the figure in terms of rules and equations. For example, what is true about the equations of parallel lines? Perpendicular lines? What features do certain shapes have? You should be able to find a rule or fact that is directly related to the fact that you are trying to prove.

EXAMPLE 17

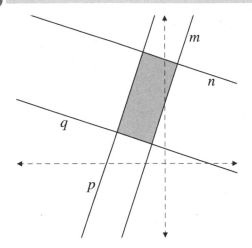

In the figure above, lines *m, n, p,* and *q* have the equations ($y = 3x + 4$), ($3y = -x + 21$), ($\frac{y}{3} = x + 4$), and ($y = -\frac{1}{3}x + 1$), respectively.

Is the shaded quadrilateral, which is bound by *m, n, p,* and *q*, a rectangle?

First, consider the definition of a rectangle—it is a figure with four sides that meet at right angles. Therefore, we can prove that the figure is a rectangle by proving that the adjacent sides meet at right angles.

One way to prove that the adjacent sides are perpendicular to each other is to identify their slopes. We have the equations for these lines, so that makes comparing the slopes easy. First, make sure that each of the equations is in the form of $y = mx + b$.

The equation for line *m* is already in standard form:

$$y = 3x + 4$$

The equation for line *n* should be put into standard form:

$$3y = -x + 21$$

$$y = \frac{-x + 21}{3} \qquad \text{Divide both sides of the equation by 3.}$$

$$y = -\frac{x}{3} + \frac{21}{3} \qquad \text{Simplify.}$$

$$y = -\frac{1}{3}x + \frac{21}{3}$$

$$y = -\frac{1}{3}x + 7$$

The equation for line p should be put into standard form:

$$\frac{y}{3} = x + 4$$

$$y = 3(x + 4) \qquad\qquad \text{Multiply both sides of the equation by 3.}$$

$$y = 3x + 12 \qquad\qquad \text{Simplify.}$$

The equation for line q is already in standard form:

$$y = -\frac{1}{3}x + 1$$

Now that the equations are in standard form, we can compare their slopes. Lines m and p both have a slope of 3, while lines n and q both have a slope of $-\frac{1}{3}$. Perpendicular lines have opposite reciprocal slopes, which is what we see here.

Line m is perpendicular to lines n and p, line n is perpendicular to m and q, and so on. Therefore, we can be sure that the figure is a rectangle, because its vertices are 90° angles.

EXAMPLE 🔒18

In the coordinate plane, a quadrilateral is bound by the lines m ($3y = 6x + 12$), n ($y = \left(\frac{1}{3}\right)x + 9$), p ($\frac{y - 9}{2} = x$), and q ($y - 2 = 5x$).

Is the quadrilateral a parallelogram?

We know that a parallelogram is a quadrilateral in which opposite sides are parallel and congruent. We can try to prove that opposite sides are parallel, or that they are congruent—either would be sufficient to prove that the quadrilateral is a parallelogram. Note that this works for parallelograms since one rule can't be true without the other; when working with other shapes, however, you may have to prove multiple facts together.

Since no coordinates were given, it will be easier to prove whether opposite sides are parallel. Put each equation into the form of $y = mx + b$; then you can compare the slopes.

Line *m*:

$$3y = 6x + 12$$

$$y = \frac{6x + 12}{3}$$ Divide both sides of the equation by 3.

$$y = 2x + 4$$ Simplify.

Line *n*:

$$y = \frac{1}{3}x + 9$$ Already in standard form.

Line *p*:

$$\frac{y - 9}{2} = x$$

$$y - 9 = 2x$$ Multiply both sides of the equation by 2.

$$y = 2x + 9$$ Add 9 to both sides of the equation.

Line *q*:

$$y - 2 = 5x$$

$$y = 5x + 2$$ Add 2 to both sides of the equation.

Compare the slopes of the lines. Lines *m* and *p* must be parallel, since they both have a slope of 2. However, lines *n* and *q* have different slopes ($\frac{1}{3}$ and 5), so they are not parallel. The figure is not a parallelogram, but rather a trapezoid.

How would you go about proving whether the opposite sides are congruent?

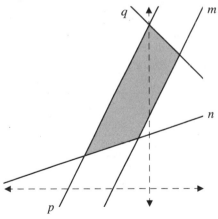

A triangle is bound by the lines ($y = x + 2$), ($y = \dfrac{2}{3}x + 2$), and ($y = \dfrac{3}{2}x - 3$).

Find the perimeter of the triangle.

To find the perimeter of the triangle, we will need to find the lengths of the sides. Since we only have equations given, a good first step would be to find the intersection points of the lines. These intersection points will be the vertices of our triangle.

Find the intersection of $y = x + 2$ and $y = \dfrac{2}{3}x + 2$. Set the expressions equal to each other:

$$x + 2 = \frac{2}{3}x + 2$$

$$x = \frac{2}{3}x \qquad \text{Subtract 2 from both sides of the equation.}$$

$$x - \frac{2}{3}x = 0 \qquad \text{Subtract } \frac{2}{3}x \text{ from both sides of the equation.}$$

$$\frac{1}{3}x = 0 \qquad \text{Simplify.}$$

$$x = 0 \qquad \text{Divide both sides of the equation by } \frac{1}{3}.$$

The x-coordinate of the intersection is 0. Plug $x = 0$ into one of the two equations to solve for y:

$$y = x + 2$$
$$y = 0 + 2$$
$$y = 2$$

Therefore, the intersection of $y = x + 2$ and $y = \dfrac{2}{3}x + 2$ is **(0, 2)**.

Can you find the remaining two intersections? Try to calculate them on your own. Then read on for the coordinates.

The intersection of ($y = x + 2$) and ($y = \dfrac{3}{2}x - 3$) is **(10, 12)**.

The intersection of ($y = \dfrac{2}{3}x + 2$) and ($y = \dfrac{3}{2}x - 3$) is **(6, 6)**.

Next, we'll need to calculate the distance between each pair of points. Use the distance formula.

$$d = \sqrt{\left(x_1 - x_2\right)^2 + \left(y_1 - y_2\right)^2}$$

Find the distance from (0, 2) to (10, 12).

$$d = \sqrt{\left(x_1 - x_2\right)^2 + \left(y_1 - y_2\right)^2}$$
$$= \sqrt{\left(10 - 0\right)^2 + \left(12 - 2\right)^2}$$
$$= \sqrt{\left(10\right)^2 + \left(10\right)^2}$$
$$= \sqrt{100 + 100}$$
$$= \sqrt{200}$$

The distance from (0, 2) to (10, 12) is $\sqrt{200}$, or ≈ 14.14.

Find the distance from (0, 2) to (6, 6).

$$d = \sqrt{\left(x_1 - x_2\right)^2 + \left(y_1 - y_2\right)^2}$$
$$= \sqrt{\left(6 - 0\right)^2 + \left(6 - 2\right)^2}$$
$$= \sqrt{\left(6\right)^2 + \left(4\right)^2}$$
$$= \sqrt{36 + 16}$$
$$= \sqrt{52}$$

The distance from (0, 2) to (6, 6) is $\sqrt{52}$, or ≈ 7.21.

Find the distance from (6, 6) to (10, 12).

$$d = \sqrt{(x_1 - x_2)^2 + (y_1 - y_2)^2}$$
$$= \sqrt{(10 - 6)^2 + (12 - 6)^2}$$
$$= \sqrt{(4)^2 + (6)^2}$$
$$= \sqrt{16 + 36}$$
$$= \sqrt{52}$$

The distance from (6, 6) to (10, 12) is $\sqrt{52}$, or ≈ 7.21.

Finally, add the three lengths to find the perimeter:

$$\sqrt{200} + \sqrt{52} + \sqrt{52}$$
$$\approx 14.14 + 7.21 + 7.21$$
$$\approx 28.56$$

19

CHAPTER 6 PRACTICE QUESTIONS

Directions: Complete the following problems as specified by each question. For extra practice after answering each question, try using an alternative method to solve the problem or check your work.

1. Line p is represented by the equation $3y + 2x = 9$. At what point will line p intersect the line that contains points $(0, 2)$ and $(5, 8)$?

2. What is the area of a triangle with vertices $(2, 1)$, $(-5, 1)$, and $(-5, -9)$?

3. On line segment JKL, JK and KL have a ratio of 1:3. If point J is at $(0, 3)$ and point K is at $(3, 7)$, what is the total distance of JKL?

4. Write the equation of a parabola with focus $(-3, 9)$ and directrix $x = 0$.

5. A parabola has an equation of $y = \left(\dfrac{1}{2}\right)(x + 3)^2 + 2$. What is the equation of the horizontal line that passes through the parabola's vertex?

6. A triangle is bound by the lines $3y + 4x = -11$, $3y - 4x = 29$, and $x = -2$. Is this a right triangle? Prove why or why not.

7. Is the triangle described in question 6 an isosceles triangle? Prove why or why not.

8. A bakery opened and quickly grew in popularity. Later, however, the popularity reached a peak and ultimately died down to the point that the bakery had to close. The bakery's success can be represented with the equation $y = -(x - \sqrt{5})^2 + 5$ with x representing the length of time in years since the bakery opened and y representing the number of customers per month, in thousands, where x and y are both positive values. After how many years did the bakery's popularity peak and what was the maximum number of customers?

9. Aiden lives 2 miles east and 3 miles north of a park. Mia lives 1 mile east and 4 miles south of the same park. Aiden and Mia want to meet exactly in the middle of their two homes. How many miles will each person have to travel, and what is the relationship between that point and the park?

SOLUTIONS TO CHAPTER 6 PRACTICE QUESTIONS

1. **(15/28, 37/14)**
    First, find the equation of the second line. The first step is to calculate the slope using the

    formula $\dfrac{y_2 - y_1}{x_2 - x_1}$. In this case, the formula is $\dfrac{8 - 2}{5 - 0} = \dfrac{6}{5}$. So far, the equation is $y = (6/5)x + b$.
    Now, calculate b by substituting the x- and y-values from one of the points: $8 = 6/5(5) + b$, then

    $8 = 6 + b$, so $b = 2$. The full equation is $y = 6/5x + 2$.

    Next, rewrite the other equation in the same form. First, $3y = -2x + 9$. Then, divide by 3 to get
    $y = (-2/3)x + 3$.

    Now, set the two equations equal to each other: $(6/5)x + 2 = -2/3x + 3$. To make the math easier, multiply the whole equation by a common denominator, which is 15. The result is $18x + 30 = -10x + 45$. Now, combine like terms to get $28x = 15$, so $x = 15/28$. Next, plug that into one of the equations to find the y-value: $y = 6/5(15/28) + 2$, then $y = 90/140 + 2 = 9/14 + 2 = 37/14$.

    The coordinate is (15/28, 37/14).

2. **35**
    First, draw the triangle to make the problem easier. You can tell that this is a right triangle, because the legs are parallel to the axes (one leg lies on the line $y = 1$, and the other leg lies on the line $x = -5$). This makes the math easier—you could use the distance formula, but in this case it's easiest to just count. The base goes from -5 to 2 on the x-axis, which is a total distance of 7. The height goes from 1 to -9 on the y-axis, for a total distance of 10. Plug those numbers into the area formula $A = (1/2)bh$ to get $A = 1/2(7)(10) = 35$.

3. **20**
    First, find the distance between (0, 3) and (3, 7). Using a right triangle, the horizontal distance is 3 and the vertical distance is 4, so the hypotenuse is 5 (3:4:5 triangle). You can also use the distance formula to get the same result. This distance is 5, and that is the first part of the 1:3 ratio, which means the other part, KL, measures 3 times as much, or 15. The total distance is $5 + 15 = 20$.

4. **$x = (-1/6)(y - 9)^2 - 3/2$**
    Use the formula for a horizontal parabola: $x = a(y - k)^2 + h$. The vertex must be halfway between the directrix and the focus. Drawing the figure can help to visualize this. The middle of 0 and -3 is -1.5, so the vertex is at $(-1.5, 9)$. The vertex is (h, k), so now the equation is $x = a(y - 9)^2 - 1.5$. Since $1/(4a)$ is the distance between the vertex and the focus (-1.5), set $1/(4a) = -1.5$, then $6a = -1$ so $a = -1/6$. Keep in mind it must be negative for the parabola to face left, which it does based on the directrix and focus. The final equation is $x = (-1/6)(y - 9)^2 - 3/2$.

5. **$y = 2$**

First find the vertex. The vertex is (h, k), so based on the equation provided it is $(-3, 2)$. The line must pass through that point. Since it is a horizontal line, it will be in the $y =$ format. The y-value of the vertex is 2, so the line is $y = 2$.

6. **No.**

In order for the triangle to be a right triangle, two sides must be perpendicular, in which case the lines' slopes are the negative reciprocals of each other. For $x = -2$, a vertical line, a perpendicular line would have to be a horizontal line, which neither of the other two are. So just check the other two to compare their slopes. First, put them into slope-intercept form. The first one is $3y + 4x = -11$, so then $3y = -4x - 11$, and $y = -4/3x - 11/3$. If the other line is perpendicular, it must have a slope of $3/4$. The other one is $3y - 4x = 29$, so then $3y = 4x + 29$ and $y = 4/3x + 29/3$. This line is not perpendicular to the other because its slope is $4/3$, not $3/4$. Another way of thinking about it is the products of the slopes must equal -1 and in this case the product is $-16/9$, not -1. Thus, this is not a right triangle.

An alternative method would be to find the vertices (the intersection points of the lines), calculate the distances between them, and then see if the distances satisfy the Pythagorean theorem. But that would be more work for this problem.

7. **Yes.**

An isosceles triangle has two equal sides. Find the distances between each pair of vertices to determine whether two of the distances are the same. First, find the vertices, which are the points at which the lines intersect. Start with $y = -4/3x - 11/3$ and $y = 4/3x + 29/3$ (as we found in question 6) and set them equal to each other: $-4/3x - 11/3 = 4/3x + 29/3$. Multiply both sides by 3: $-4x - 11 = 4x + 29$. Combine like terms: $-40 = 8x$, so $x = -5$. Plug that into one of the equations: $y = -4/3(-5) - 11/3$, then $y = 20/3 - 11/3 = 9/3 = 3$. One point is $(-5, 3)$.

Now, combine $y = -4/3x - 11/3$ and $x = -2$. The x-value is -2, so plug that in to find y: $y = -4/3(-2) - 11/3 = 8/3 - 11/3 = -3/3 = -1$. A second vertex is $(-2, -1)$. Next, do the same to find the last vertex with $y = 4/3x + 29/3$ and $x = -2$. $y = 4/3(-2) + 29/3 = -8/3 + 29/3 = 21/3 = 7$. The last vertex is $(-2, 7)$.

Next, use the distance formula or right triangles to find the distances between the points $(-5, 3)$, $(-2, -1)$, and $(-2, 7)$. For $(-5, 3)$ and $(-2, -1)$, the difference between the x-values is 3 and between the y-values is 4, so this is a 3:4:5 triangle. Their distance is 5. For $(-2, -1)$ and $(-2, 7)$, the distance is just the difference between the y-values, which is 8. For $(-2, 7)$ and $(-5, 3)$, the difference between the x-values is 3, between the y-values is 4, so again this is a 3:4:5 and their distance is 5. Thus, this is an isosceles triangle because two sides have an equal distance of 5.

8. **About 2.2 years and 5,000 customers.**

Any time you're asked to find the maximum or minimum value for a parabola, you're being asked to find the vertex. The equation is in vertex form, so the vertex is (h, k), or $(\sqrt{5}, 5)$ (approximately $(2.2, 5)$). The x represents the amount of time in years, so that is 2.2 years, and the y represents the number of customers in thousands, so 5,000 customers.

9. **About 3.5 miles per person; 3/2 mile east and 1/2 mile south of the park.**
   Think of the park like the origin on a coordinate grid. Aiden's point is (2, 3) and Mia's point is

   (1, −4). The question is asking for the midpoint, which is found by taking the average of the

   x-values and the average of the y-values. Here it is: $\left(\dfrac{2+1}{2}\right),\left(\dfrac{3+(-4)}{2}\right)$, or (3/2, −1/2). This

   means that the meeting spot is 3/2 mile east of the park and 1/2 mile south of the park. Now

   determine how far each person has to travel. This involves the distance between the two points,

   which can be found using the distance formula or a right triangle. The difference between the

   x-values is 1 and the difference between the y-values is 7, so the formula is $1^2 + 7^2 = c^2$ or

   $50 = c^2$, so $c = \sqrt{50}$, which is about 7.1. They will each travel half that distance, so about 3.5

   miles per person.

REFLECT

**Congratulations on completing Chapter 6!
Here's what we just covered.
Rate your confidence in your ability to:**

- Understand the standard form for the equation of a line

① ② ③ ④ ⑤

- Find an equation from a line, and plot a line from its equation

① ② ③ ④ ⑤

- Find the intersection of two lines

① ② ③ ④ ⑤

- Calculate the distance or midpoint between two coordinates

① ② ③ ④ ⑤

- Divide a segment to a specified ratio (e.g., 2:3 or 1:5)

① ② ③ ④ ⑤

- Find the vertex, roots, and axis of symmetry of a parabola

① ② ③ ④ ⑤

- Recognize the standard form and vertex form for the equation of a parabola

① ② ③ ④ ⑤

- Find the focus and directrix of a parabola from its equation

① ② ③ ④ ⑤

- Find the equation of a parabola given its zeroes, or its focus and directrix

① ② ③ ④ ⑤

- Prove geometric facts and theorems using algebra

① ② ③ ④ ⑤

If you rated any of these topics lower than you'd like, consider reviewing the corresponding lesson before moving on, especially if you found yourself unable to correctly answer one of the related end-of-chapter questions.

Access your online student tools for a handy, printable list of Key Points for this chapter. These can be helpful for retaining what you've learned as you continue to explore these topics.

Chapter 7
Circles

GOALS By the end of this chapter, you will be able to:

- Understand tangent lines, secant lines, and chords, and the relationships among these and other parts of a circle

- Calculate the length of an arc, or the area of a sector of a circle

- Solve problems using the ratios of arc length : circumference, sector area : total area, and central angle : total angle

- Understand the relationships between inscribed and central angles

- Solve problems with concentric circles

- Understand radians, and convert between radians and degrees

Lesson 7.1
Advanced Circles

SECANTS AND CHORDS

In Lesson 5.1, we reviewed radius, diameter, circumference, and area. In this chapter, you'll build on your knowledge of circles, and the relationships among different parts of a circle.

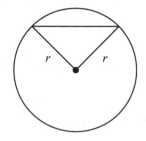

If we have a triangle inscribed in a circle as shown above, with two vertices on the circle's circumference and one vertex at its center, then the triangle is certainly isosceles. We know this because *all radii in a circle are equal*. In the figure, two of the triangle's sides are radii of the circle; therefore, they are congruent.

A circle has infinitely many triangles that can be formed in this manner.

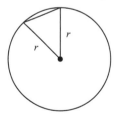
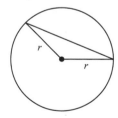
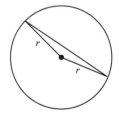

The line segment that connects two points on a circle is called a **chord**. If that segment is extended to become a line, that line is called a **secant**. If a chord passes through the center of a circle, then it is a **diameter**.

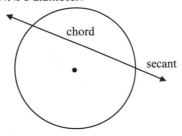

If a radius is perpendicular to a chord, then it bisects the chord. The converse is also true—if a radius bisects a chord, then it is perpendicular to the chord.

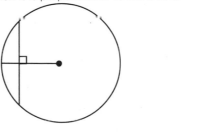

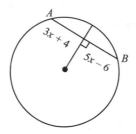

Find the length of *AB*.

In this figure, the two halves of *AB* are congruent, since they are bisected by a perpendicular radius. If they are congruent, that means that we can set them equal to each other:

$$3x + 4 = 5x - 6$$
$$3x + 10 = 5x \qquad \text{Add 6 to both sides of the equation.}$$
$$10 = 2x \qquad \text{Subtract } 2x \text{ from both sides of the equation.}$$
$$5 = x$$

We've concluded that $x = 5$. But don't forget to solve for *AB*! Plug in 5 for *x* in both expressions, and add them together.

$$3x + 4 + 5x - 6$$
$$= 3(5) + 4 + 5(5) - 6$$
$$= 15 + 4 + 25 - 6$$
$$= 19 + 25 - 6$$
$$= 44 - 6$$
$$= 38$$

The length of *AB* is 38.

TANGENTS

A line **tangent** to a circle intersects the circle at exactly one point. The intersection point is known as the **point of tangency**. The radius intersects the point of tangency at a 90° angle.

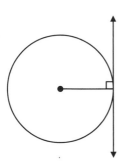 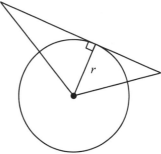

Consider a triangle that is tangent to a circle, as shown in the right-hand figure above. One vertex is at the center of the circle, and the other two vertices are on the tangent line. The triangle is not necessarily isosceles; however, we do know a little bit more than that. If we connect a radius at the point of tangency, the radius would be perpendicular to the tangent line, and in fact, the radius would serve as an **altitude** of the triangle.

Here is how you may see chords on the ACT.

In the circle below, radius $\overline{OP}$ is 10 inches long, $\angle LOP$ is 60°, and $\overline{OP}$ is perpendicular to chord $\overline{LN}$ at M. How many inches long is $\overline{LN}$?

A. $10\sqrt{3}$

B. $3\sqrt{15}$

C. 10

D. $5\sqrt{3}$

E. 5

A circle has infinitely many triangles that can be formed in this manner.

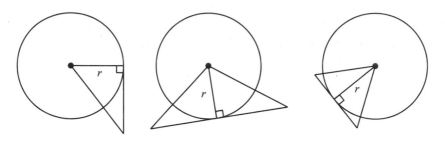

EXAMPLE 2

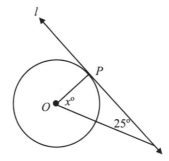

In the figure above, line *l* is tangent to the circle with center *O* at point *P*. What is the value of *x*?

If line *l* is tangent to the circle, then we know that line *l* is perpendicular to the radius shown. That means that there is only one unknown angle, *x*, in the triangle. To solve for *x*, subtract the two known angles from 180°.

$$x = 180° - 90° - 25°$$
$$x = 90° - 25°$$
$$x = 65°$$

If two tangent lines intersect, then each has the same length from the intersection point to the point of tangency.

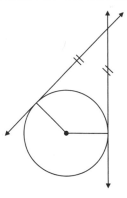

Start by labeling the length of $\overline{OP}$ and the measurement of $\angle LOP$ on the figure. $\overline{LO}$ is another radius of the circle, so label that with a length of 10 inches as well. $\triangle LMO$ is a 30-60-90 triangle, with $\angle MLO$ equal to 30°. $\overline{LO}$ is opposite the 90° angle, so $\overline{LO}$ is the $2x$ side. If $2x = 10$, then $x = 5$. MO is opposite the 30° angle, so $\overline{MO}$ is the x side and is equal to 5. $\overline{LM}$ is opposite the 60° angle, so it is the $x\sqrt{3}$ side and is equal to $5\sqrt{3}$. This is part of the length of $\overline{LN}$, but the length of $\overline{MN}$ still needs to be found. Draw in the radius connecting points N and O to create another triangle. Because $\overline{LO}$ and $\overline{NO}$ are the same length and $\overline{LN}$ is perdendicular to $\overline{OP}$, the new triangle is congruent to $\triangle LMO$. Therefore, $\overline{MN}$ is equal to $\overline{LM}$, and $\overline{LN}$ equals $5\sqrt{3} + 5\sqrt{3}$ or $10\sqrt{3}$. Choice (A) is the credited response.

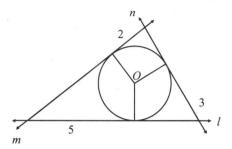

In the figure above, lines *l*, *m*, and *n* are each tangent to the circle with center *O*. What is the perimeter of the triangle formed by lines *l*, *m*, and *n*?

From the rule above, we know that two intersecting tangent lines will form congruent segments. Find the segments that are congruent in this figure.

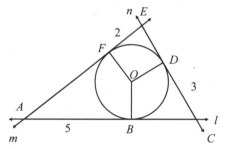

For the purposes of this exercise, we've labeled the vertices and tangent points in the figure. Segments *AB* and *AF* are intersecting tangent segments, so we know that they are congruent. Therefore, *AF* has a length of 5.

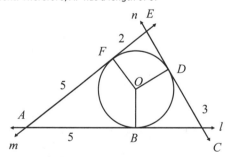

For the same reasons, we know that *BC* is congruent to *CD*, and that *DE* is congruent to *EF*. Therefore, *CD* has a length of 3 and *EF* has a length of 2.

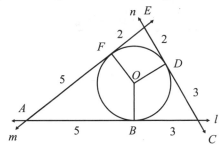

Add up all the lengths to find the perimeter:

$$2 + 2 + 3 + 3 + 5 + 5 = 20$$

The triangle's perimeter is 20.

Lesson 7.2
Arcs and Sectors

ARCS

An **arc** is a portion of the circumference of a circle. Arc and circumference are both measured in units of length—in other words, they measure the distance along the edge of a curve.

The term *arc* can also refer to the angle formed by the endpoints of the arc and the center of the circle. Whenever an arc is defined in a circle, there are actually two arcs— the **major arc** and **minor arc**. A major arc has an angle greater than 180°, while a minor arc has an angle less than 180°. In the figure below, the major arc has an angle of 265°, and the minor arc has an angle of 95°.

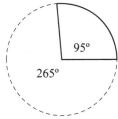

EXAMPLE 4

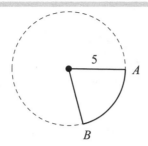

In the figure above, minor arc **AB** measures $\frac{1}{5}$ the circumference of the circle. **What is the measure of arc *AB*?**

The radius of the circle is given as 5, so the circumference is 10π (circumference = $2\pi r$). It's also stated that the arc measures $\frac{1}{5}$ of the circumference of the circle. Therefore, the arc must be equal to $10\pi \times \frac{1}{5}$, or 2π.

A **semicircle** is exactly half of a circle, and its arc measures 180°. A semicircle arc is not considered "major" or "minor."

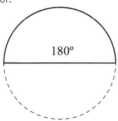

EXAMPLE 5

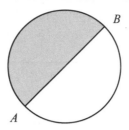

In the figure above, semicircle *AB* has an area of 18π. What is the radius of the circle?

Note that the area given was for the semicircle, not the full circle. Don't let that confuse you, and don't worry about deriving a formula to solve for the radius of a semicircle. Right away, you should multiply that area by 2 to get the full area of the circle, which is 36π. Then, you can use the area of a circle to solve for the radius.

$$A = \pi r^2$$
$$36\pi = \pi r^2$$
$$36 = r^2 \qquad \text{Divide both sides of the equation by } \pi.$$
$$6 = r \qquad \text{Take the square root of both sides of the equation.}$$

The radius of the circle is 6.

5

Here is how you may see semicircles on the ACT.

An equilateral triangle and 2 semicircles have dimensions as shown in the figure below. What is the perimeter, in inches, of the figure?

A. $3 + 3\pi$
B. $6 + 6\pi$
C. $6 + 12\pi$
D. $8 + 6\pi$
E. $18 + 12\pi$

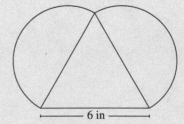

An **intercepted arc** is an arc that is defined by two lines or segments intersecting its endpoints. It's no different from any other arc, but you should be aware that an intercepted arc is considered to be the arc *between* two line segments. In the figure below, the intercepted arcs are shown in bold. Sometimes, there are two in the same circle.

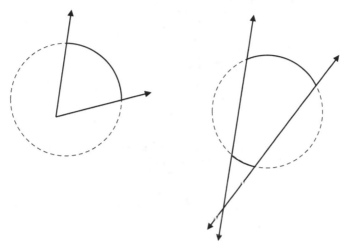

SECTORS

A **sector** is a portion of the area of a circle, bound by two radii and their intercepted arc. In other words, it's a "wedge" or "slice" of the circle. Sectors are measured in terms of area.

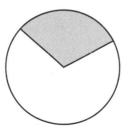

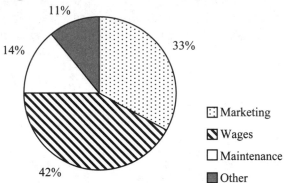

11%

14%

33%

42%

:::: Marketing

N Wages

☐ Maintenance

■ Other

The pie chart above is printed on a poster for the annual executive meeting at Krazy Town. If the area of the entire circular pie chart is 144π square inches, what is the area of the sector labeled "Maintenance"?

The given information states that the area of the circle is 144π, and the sector labeled "Maintenance" is 14% of the circle. To find the area of the sector, multiply the total area by the percentage given for the sector.

$$144\pi \times .14 \qquad\qquad \text{.14 is the decimal equivalent of 14\%.}$$
$$= 20.16 \times \pi$$
$$\approx 63.33$$

The area of the sector is approximately 63.33 square inches.

The perimeter of the figure consists of one side of the equilateral triangle and the arc length of two semicircles. You can immediately eliminate (A), (D), and (E) because the length of one side of the equilateral triangle is 6 inches. Both semicircles have a diameter of 6 and given $C = d\pi = 6\pi$, each semicircle has an arc length of 3π. With the exposed side of the triangle, the perimeter of the figure should be $P = 6 + 3\pi + 3\pi = 6 + 6\pi$. If you selected (C), you may have found the circumferences of two full circles rather than two semicircles. The correct answer is (B).

ACT A

INSCRIBED AND CENTRAL ANGLES

An angle whose vertex is at the center of the circle is called a **central angle**. An angle whose vertex is on the circumference of the circle is called an **inscribed angle**.

 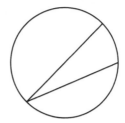

The measure of an inscribed angle is always $\frac{1}{2}$ the measure of the corresponding central angle. In this context, "corresponding" means that both pairs of legs intersect the same two points on the circle.

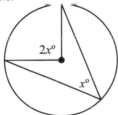

EXAMPLE 7

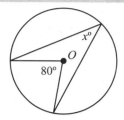

In the figure above, _O_ is the center of the circle. What is the measure of _x_?

From the rule above, we know that the inscribed angle is always $\frac{1}{2}$ the measure of the corresponding central angle. Therefore, the inscribed angle is 40° (= 80° × $\frac{1}{2}$).

EXAMPLE 8

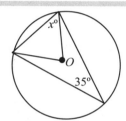

In the figure above, _O_ is the center of the circle. What is the measure of _x_?

From the rule above, we know that the inscribed angle is always $\frac{1}{2}$ the measure of the corresponding central angle. Therefore, the central angle must be 70°.

To solve for _x_, it's necessary to recognize that two sides of the triangle are radii of the circle. This means that they must be congruent sides, since all radii in a circle are congruent. Therefore, this is an isosceles triangle, and the two base angles are congruent.

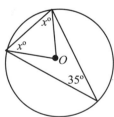

Since these two unknown angles are congruent, we can make an equation to solve:

$180° = 70° + 2x°$
$110° = 2x°$ Subtract 70° from both sides of the equation.
$55° = x$ Divide both sides of the equation by 2.

The measure of x is 55°.

CONCENTRIC CIRCLES

Concentric circles are circles that share the same central point.

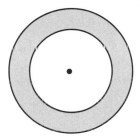

The distance between two concentric circles is the same all the way around. The "ring" or "donut" shape formed by two concentric circles is called an **annulus**.

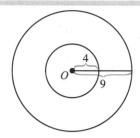

In the figure above, the two circles are concentric about point O. What is the area of the annulus?

To find the area of an annulus, first find the areas of the two circles.

Smaller circle:

$$A = \pi r^2$$
$$= \pi(4)^2$$
$$= 16\pi$$

Larger circle:

$$A = \pi r^2$$
$$= \pi(9)^2$$
$$= 81\pi$$

The area of the annulus is the difference between the areas of the two circles.

$$= 81\pi - 16\pi$$
$$= 65\pi$$

The area of the annulus is 65π.

Lesson 7.3
Slices of Pi

In this lesson, you will learn how to calculate an arc/sector based on information about a circle, and vice versa. You will also learn how to use radians, which are units of angle measure.

Normally, the easiest way to think about these problems is in terms of basic fractions—that is, **part/whole** relationships. Imagine that you have a pizza with 8 congruent slices. If you eat 1 of the slices, then you know that you ate $\frac{1}{8}$ of the total pizza.

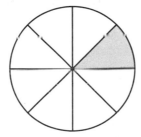

Now, if you wanted to know the degree measure of the missing pieces of pizza, that is also fairly straightforward. Since we know that the total degree measure of a circle is always 360°, we can use a proportion to find the angle of the sector. This is a part/whole relationship—the measure of the arc, compared with the measure of the full circle.

$$\frac{1}{8} = \frac{x}{360°}$$

$1 \times 360° = 8 \times x$ \qquad Cross-multiply.

$360° = 8 \times x$

$\frac{360°}{8} = x$ \qquad Divide both sides of the equation by 8.

$45° = x$

Therefore, the total degree measure of the missing pieces is 45°.

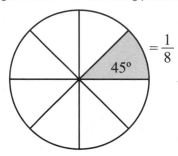

What if you wanted to know the total area of the 1 slice you ate? To do that, you would use another part/whole relationship—the area of the slice, compared with the total area of the pizza. So, you would need some information about the size of the pizza. Let's say that you're told that the radius of the circle is 12 inches. That would be enough to tell you the area of the full pizza:

$$r = 12$$
$$A = \pi r^2$$
$$= \pi(12)^2$$
$$= 144\pi$$

Therefore, the area of the pizza is 144π square inches. Now, we can use a proportion to find the area of the sector.

$$\frac{1}{8} = \frac{x}{144\pi}$$

$1 \times 144\pi = 8 \times x$ Cross-multiply.

$144\pi = 8 \times x$

$\dfrac{144\pi}{8} = x$ Divide both sides of the equation by 8.

$18\pi = x$ Simplify.

Therefore, the total area of the missing pieces is 18π square inches.

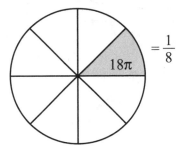

Let's calculate the length of the arc (because obviously, the crust is the best part). To do that, we'll use another part/whole relationship—the length of the arc from the slices, compared with the total circumference of the pizza. We were told that the radius of the circle is 12 inches. With that, we can calculate the circumference of the pizza:

$$r = 12$$
$$C = 2\pi r$$
$$= 2\pi(12)$$
$$= 24\pi \qquad \text{Simplify.}$$

Therefore, the circumference of the pizza is 24π inches. Now, we can use a proportion to find the length of the arc.

$$\frac{1}{8} = \frac{x}{24\pi}$$

$$1 \times 24\pi = 8 \times x \qquad \text{Cross-multiply.}$$

$$24\pi = 8 \times x$$

$$\frac{24\pi}{8} = x \qquad \text{Divide both sides of the equation by 8.}$$

$$3\pi = x \qquad \text{Simplify.}$$

Therefore, the length of the arc for the missing pieces is 3π inches.

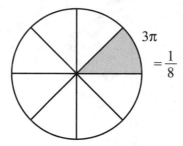

To solve a problem with arc or sectors, remember that all of the following are proportional:

Parts of a Circle

$$\frac{\text{part}}{\text{whole}} = \frac{\text{central angle}}{360°} = \frac{\text{arc length}}{\text{total circumference}} = \frac{\text{sector area}}{\text{total area}}$$

You can solve for **all** of these parts if you have only two: something that tells you the size of the full circle, and something that tells you the size of the wedge.

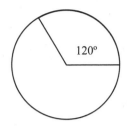

In the figure above, the circle has a radius of 9. What is the area of the sector with a central angle of 120°?

With the radius given, you can solve for the area of the circle:

$$A = \pi r^2$$
$$= \pi(9)^2$$
$$= 81\pi$$

You also know the angle measure of the sector, so to set up a proportion:

$$\frac{120°}{360°} = \frac{x}{81\pi}$$

$$\frac{1}{3} = \frac{x}{81\pi} \qquad \text{Simplify.}$$

$$1 \times 81\pi = 3 \times x \qquad \text{Cross-multiply.}$$

$$81\pi = 3 \times x$$

$$\frac{81\pi}{3} = x \qquad \text{Divide both sides of the equation by 3.}$$

$$27\pi = x \qquad \text{Simplify.}$$

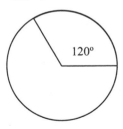 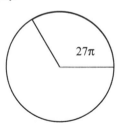

The area of the sector is 27π.

EXAMPLE 11

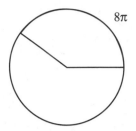

In the figure above, the circle has an area of 100π. What is the central angle of the arc with a length of 8π?

In this example, we'll need to make a proportion that uses the fraction of arc length/circumference. But we weren't given circumference, we were given area. No problem—we can use the area to solve for the radius, and then solve for the circumference.

Remember CARD from Chapter 5; you can solve for circumference, area, radius, *and* diameter if you have just one of those pieces of information given.

$$A = \pi r^2$$
$$100\pi = \pi r^2$$
$$100 = r^2 \qquad \text{Divide both sides of the equation by } \pi.$$
$$10 = r \qquad \text{Take the square root of both sides of the equation.}$$

With the radius known, we can solve for the circumference:

$$C = 2\pi r$$
$$C = 2\pi(10)$$
$$C = 20\pi \qquad \text{Simplify.}$$

Now, set up a proportion to solve for the central angle of the arc. On one side of the equation, use the given arc length with circumference.

$$\frac{8\pi}{20\pi} = \frac{x}{360°}$$

$$\frac{8}{20} = \frac{x}{360°} \qquad \text{Simplify.}$$

$$\frac{2}{5} = \frac{x}{360°} \qquad \text{Simplify.}$$

$$2 \times 360° = 5 \times x \qquad \text{Cross-multiply.}$$

$720° = 5 \times x$

$$\frac{720°}{5} = x$$ Divide both sides of the equation by 5.

$144° = x$

Therefore, the degree measure of the arc is 144°.

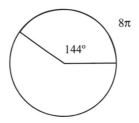

Here is how you may see arcs on the SAT.

Points A and B lie on circle O as shown below. $\angle BOA$ is 45°. If the area of circle O is 64π, what is the length of minor arc AB?

A. 2π
B. 4π
C. 8π
D. 16π

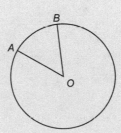

Pi Practice

Try completing the table below. Each row gives two pieces of information—something about the whole circle, and something about the arc or sector. For example, if area and central angle are given in a row, try solving for sector area first, then circumference and arc length. The first row has been completed.

Circumference = $2\pi r$	Area = πr^2	Central Angle	Arc Length	Sector Area
12π	36π	60°	2π	6π
16π		270°		
	25π		6π	
6π			4π	
	16π	90°		
	100π			80π
24π				60π
18π		160°		

RADIANS

The **radian** is a unit of angle measure. One radian represents an angle in which the arc length is exactly equal to the radius of a circle. It is usually described in terms of a circle with a radius of 1. In a circle with a radius of 1, one radian corresponds to an arc with a length of 1.

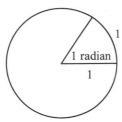

A circle has precisely 2π (≈ 6.283) radians, a measure equal to 360°.

Angles measured in radians are best expressed as fractions of 2π. For example, a half circle has an angle measure of π radians ($=\frac{1}{2}$ of 2π), and a quarter circle has an angle measure of $\frac{\pi}{2}$ radians ($=\frac{1}{4}$ of 2π).

Mathematicians like to use radians, because it is a unit based on a "real" relationship in a circle—the relationship between the arc and the radius. Degrees, by contrast, are actually arbitrary units. The reason we use 360° is because 360 has a lot of factors, which makes division easy.

If you see an angle measure expressed in terms of π, you can be sure that you're dealing with radians. Sometimes, the abbreviation "r" is used, but it is not necessary.

You know the area is 64π, so you can solve for the radius, which is 8. That makes the circumference 16π. Since $\angle BOA$ is 45°, which is $\frac{1}{8}$ of the 360° of a circle, the length of arc AB will be $\frac{1}{8}$ of the circumference. $\frac{1}{8}$ of 16π is 2π. The correct answer is (A).

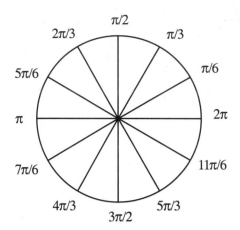

If the figure is confusing, it may help you to write down the unreduced form of each fraction, as a multiple of $\frac{1}{6}$. For example, $\frac{\pi}{3}$ is equal to $\frac{2\pi}{6}$, and so on. You'll see that the values are sequential; $\frac{\pi}{6}, \frac{2\pi}{6}, \frac{3\pi}{6}$, etc.

To convert radians to degrees, or vice versa, use the following proportion:

Converting Radians and Degrees

$$\frac{radians}{\pi} = \frac{degrees}{180}$$

EXAMPLE 12

What is the equivalent of 216 degrees, as expressed in radians?

To convert degrees to radians, use the proportion above.

$$\frac{x}{\pi} = \frac{216}{180}$$

$$\frac{x}{\pi} = \frac{6}{5} \qquad \qquad \text{Simplify.}$$

$$x = \frac{6}{5}\pi \qquad \qquad \text{Multiply both sides of the equation by } \pi.$$

The angle measure equivalent to 216 degrees is $\frac{6\pi}{5}$ radians.

Here is how you may see radians on the ACT.

Which of the following degree measures is equivalent to 2.25π radians?

A. 101.25

B. 202.5

C. 405

D. 810

E. 1,620

Radians Practice

Try the following conversions. Answers are at the end of this chapter.

Degrees	Radians
30	
	$\dfrac{\pi}{4}$
60	
	$\dfrac{2\pi}{5}$
90	
120	
	π
	$\dfrac{3\pi}{2}$

ARCS AND SECTORS USING RADIANS

To solve a problem with arc or sector and radians, use the following proportions:

Parts of a Circle

$$\frac{\text{part}}{\text{whole}} = \frac{\text{central angle}}{2\pi} = \frac{\text{arc length}}{\text{total circumference}} = \frac{\text{sector area}}{\text{total area}}$$

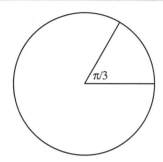

In the figure above, the circle has a circumference of 36π. What is the length of the arc with a central angle of $\frac{\pi}{3}$?

With the circumference and angle measure given, you can set up a proportion:

$$\frac{\frac{\pi}{3}}{2\pi} = \frac{x}{36\pi}$$

$$\frac{\pi}{3 \times 2\pi} = \frac{x}{36\pi} \qquad \text{Simplify.}$$

$$\frac{1}{6} = \frac{x}{36\pi} \qquad \text{Simplify.}$$

$$1 \times 36\pi = 6 \times x \qquad \text{Cross-multiply.}$$

$$36\pi = 6 \times x$$

Don't be afraid to begin with nested fractions in your proportion. You can simplify in the next step. It's better to get everything down in the format you know, first.

To change 2.25π radians into degrees, multiply by $\frac{180}{\pi}$ to get 405°. Choices (A) and (B) are fractions of 405. Choices (C) and (E) are multiples. The correct answer is (C).

$$\frac{36\pi}{6} = x$$ Divide both sides of the equation by 6.

$$6\pi = x$$ Simplify.

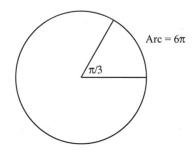

The length of the arc is 6π.

13

EXAMPLE 14

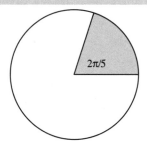

In the figure above, the circle has a radius of 5. What is the area of the sector with a central angle of $\dfrac{2\pi}{5}$?

With the radius given, you can solve for the area of the circle:

$$A = \pi r^2$$
$$= \pi(5)^2$$
$$= 25\pi$$

You also know the angle measure of the sector, so set up a proportion:

$$\frac{\frac{2\pi}{5}}{2\pi} = \frac{x}{25\pi}$$

$$\frac{2\pi}{5 \times 2\pi} = \frac{x}{25\pi} \qquad \text{Simplify.}$$

$$\frac{1}{5} = \frac{x}{25\pi} \qquad \text{Simplify.}$$

$$1 \times 25\pi = 5 \times x \qquad \text{Cross-multiply.}$$

$$25\pi = 5 \times x$$

$$\frac{25\pi}{5} = x \qquad \text{Divide both sides of the equation by 5.}$$

$$5\pi = x \qquad \text{Simplify.}$$

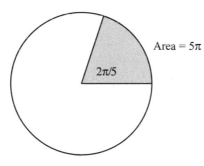

The area of the sector is 5π.

ANSWERS TO CHAPTER 7 EXERCISES

Pi Practice (Page 322)

Circumference $= 2\pi r$	Area $= \pi r^2$	Central Angle	Arc Length	Sector Area
12π	36π	$60°$	2π	6π
16π	64π	$270°$	12π	48π
10π	25π	$216°$	6π	15π
6π	9π	$240°$	4π	6π
8π	16π	$90°$	2π	4π
20π	100π	$288°$	16π	80π
24π	144π	$150°$	10π	60π
18π	81π	$160°$	8π	36π

Radians Practice (Page 326)

Degrees	Radians
30	$\dfrac{\pi}{6}$
45	$\dfrac{\pi}{4}$
60	$\dfrac{\pi}{3}$
72	$\dfrac{2\pi}{5}$
90	$\dfrac{\pi}{2}$
120	$\dfrac{2\pi}{3}$
180	π
270	$\dfrac{3\pi}{2}$

CHAPTER 7 PRACTICE QUESTIONS

Directions: Complete the following problems as specified by each question. For extra practice after answering each question, try using an alternative method to solve the problem or check your work.

1. An angle measures 4 radians. What is the measure of the angle in degrees?

2. If the area of a circle is 9, what is the circumference?

3. In the figure below, what is the area of the circle?

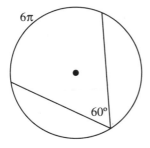

4. The circumference of the circle below is 12π and the center is O. If the area of the sector formed by $\overline{AO}$, $\overline{BO}$, and minor arc AB is 4π, what is the measure of $\angle AOB$ in radians?

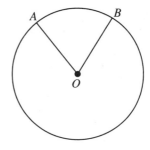

5. In the figure below, if the radius of the circle with center O is 4, what is the area of triangle PQR?

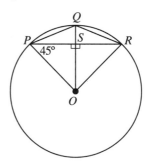

6. If the area of the annulus between two concentric circles is 64π and the ratio between the radii of the two circles is 5:3, what is the area of the larger circle?

7. In the figure below, both circles have center O and $\angle AOC$ is $\dfrac{\pi}{4}$ radians. If two circles have circumferences 16π and 8π, respectively, what is the area of the shaded region?

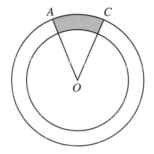

8. In the figure below, if the radius of the circle with center O is 1, what is the area of quadrilateral $ABCO$?

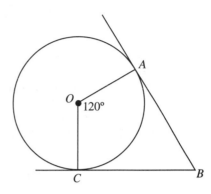

SOLUTIONS TO CHAPTER 7 PRACTICE QUESTIONS

1. **$\dfrac{720}{\pi}$**

   To convert the angle from radians to degrees, use the proportion $\dfrac{\pi \ radians}{180\,°} = \dfrac{4 \ radians}{x°}$.

   Cross-multiply to get $720 = \pi x$. Divide both sides by π to get $\dfrac{720}{\pi}$.

2. **$6\sqrt{\pi}$**

   The formula for area of a circle is $A = \pi r^2$. Plug in the area of the circle to get $9 = \pi r^2$. Divide both

   sides by π to get $\dfrac{9}{\pi} = r^2$. Take the square root of both sides to get $r = \sqrt{\dfrac{9}{\pi}} = \dfrac{\sqrt{9}}{\sqrt{\pi}} = \dfrac{3}{\sqrt{\pi}}$. Plug

   this into the circumference formula to get $C = 2\pi r = 2\pi\left(\dfrac{3}{\sqrt{\pi}}\right) = \dfrac{6\pi}{\sqrt{\pi}} = 6\sqrt{\pi}$.

   Reminder: If you have an irrational number in the denominator, you should multiply the

   fraction by a value of 1 (in this case, multiply $\dfrac{6\pi}{\sqrt{\pi}}$ by $\dfrac{\sqrt{\pi}}{\sqrt{\pi}}$.) This is known as "rationalizing" the

   denominator.

3. **81π**

   The question asks for the area of the circle, which is determined by the formula $A = \pi r^2$. The

   figure provides an inscribed angle and the length of the intercepted arc. Anytime an arc is

   mentioned in a geometry question, write the arc formula: $\dfrac{arc}{circumference} = \dfrac{angle}{360}$. The measure

   of the inscribed angle is half the measure of the central angle, so the central angle is 120°. Plug

   this angle and the arc length into the formula to get $\dfrac{6\pi}{C} = \dfrac{120}{360}$. Reduce the fraction on the

   right side to get $\dfrac{6\pi}{C} = \dfrac{1}{3}$. Cross-multiply to get $C = 18\pi$. Since $C = 2\pi r$, $18\pi = 2\pi r$. Divide both

   sides by 2π to get $r = 9$. Plug this into the area formula to get $A = \pi r^2 = \pi(9)^2 = 81\pi$.

4. **$\dfrac{2\pi}{9}$**

   The question mentions sector area, so set up a proportion: ($\dfrac{area \ sector}{area \ circle} = \dfrac{angle}{2\pi}$). Note that

   we're using radians here instead of degrees. The area of the sector is 4π. The question asks for the angle, so find the area. The question says that the circumference is 12π. The formula for circumference is $C = 2\pi r$, so $12\pi = 2\pi r$. Divide both sides by 2π to get $r = 6$. Plug this into the area formula, $A = \pi r^2$ to get $A = \pi(6)^2 = 36\pi$. Plug this into the sector formula to get $\dfrac{4\pi}{36\pi} = \dfrac{x}{2\pi}$. Reduce the left side of the equation to get $\dfrac{1}{9} = \dfrac{x}{2\pi}$. Cross-multiply to get $9x = 2\pi$. Divide both sides by 9 to get $x = \dfrac{2\pi}{9}$.

5. **$8\sqrt{2} - 8$**

Since the radius is 4, mark off any radius as having a length of 4: $OP = OQ = OR = 4$. Notice that triangle OPS is a right triangle with a 45° angle. Since there are 180° in a triangle, m$\angle POS$ must be $180° - 90° - 45° = 45°$, making triangle OPS a 45-45-90 right triangle. The sides of a 45-45-90 right triangle are in a ratio of $1:1:\sqrt{2}$. The side opposite the right angle, $\overline{OP}$, has a length of 4. To get the length of the other two sides, set up the proportion $\dfrac{1}{\sqrt{2}} = \dfrac{x}{4}$. Cross-multiply to get $x\sqrt{2} = 4$. Divide both sides by $\sqrt{2}$ to get $x = \dfrac{4}{\sqrt{2}}$. Rationalize the denominator by multiplying by $\dfrac{\sqrt{2}}{\sqrt{2}}$ to get $x = \dfrac{4}{\sqrt{2}} \times \dfrac{\sqrt{2}}{\sqrt{2}} = \dfrac{4\sqrt{2}}{2} = 2\sqrt{2}$. Therefore, $PS = OS = 2\sqrt{2}$. Since $\overline{OQ}$ is a radius that is perpendicular to chord $\overline{PR}$, $\overline{OQ}$ bisects $\overline{PR}$, so $PS = SR = 2\sqrt{2}$. Chord $\overline{PR}$ is the base of triangle PQR. The height is $\overline{QS}$. To get QS, subtract OS from OQ to get $QS = 4 - 2\sqrt{2}$. Plug these into the area formula to get $A = \dfrac{1}{2}bh = \dfrac{1}{2}(4\sqrt{2})(4 - 2\sqrt{2}) = (2\sqrt{2})(4 - 2\sqrt{2}) = 8\sqrt{2} - (2\sqrt{2})^2 = 8\sqrt{2} - 8$.

6. **100π**

The radii of the two circles have a radius of 5:3, so let their values equal $5x$ and $3x$, respectively. The area of the annulus is the difference between the areas of the two circles. Since the radius of the larger circle is $5x$, the area is $A = \pi r^2 = \pi(5x)^2 = 25\pi x^2$. Since the radius of the smaller circle is $3x$, the area is $A = \pi r^2 = \pi(3x)^2 = 9\pi x^2$. The area of the annulus is the difference between these two, which is $25\pi x^2 - 9\pi x^2 = 16\pi x^2$. The question says that this is 64π, so $64\pi = 16\pi x^2$. Divide both sides by 16π to get $4 = x^2$. Take the square root of both sides to get $x = 2$. The question asks for the area of the larger circle. Since $x = 2$, the radius is $5x = 5(2) = 10$. The area is $A = \pi r^2 = \pi(10)^2 = 100\pi$. Alternatively, plug $x = 2$ into the previous derived expression for the area of the larger circle, $25\pi x^2$, to get $25\pi(2)^2 = 25\pi(4) = 100\pi$.

7. **6π**

The shaded region is the difference between the two sectors. Determine the area of each sector

and find the difference. Start with the larger circle, which has a circumference of 16π. Sector

area directly relates to circle area rather than circumference, so use the circumference to get the

area. The circumference formula is $C = 2\pi r$, so $16\pi = 2\pi r$. Divide both sides by 2π to get $r = 8$.

Plug this into the area formula to get $A = \pi r^2 = \pi(8)^2 = 64\pi$. Plug this into the radians version of

the sector formula, $\dfrac{area\ sector}{area\ circle} = \dfrac{angle}{2\pi}$, to get $\dfrac{x}{64\pi} = \dfrac{\frac{\pi}{4}}{2\pi}$. Cross-multiply to get $16\pi^2 = 2\pi x$.

Divide both sides by 2π to get $x = 8\pi$. Follow a similar method to get the area of the smaller arc.

Since the circumference is 8π, $8\pi = 2\pi r$, and $r = 4$. Therefore, $A = \pi r^2 = \pi(4)^2 = 16\pi$. Plug this into

the sector formula to get $\dfrac{y}{16\pi} = \dfrac{\frac{\pi}{4}}{2\pi}$. Cross-multiply to get $4\pi^2 = 2\pi y$. Divide both sides by 2π

to get $y = 2\pi$. Subtract the two sectors to get $8\pi - 2\pi = 6\pi$.

8. **$\sqrt{3}$**

There's no formula for the area of a quadrilateral that isn't a parallelogram or a trapezoid, so

break this into figures that have a formula: in this case, two triangles. Since $\overline{CB}$ and $\overline{AB}$ are

tangent to the circle, they form right angles with the radius. Since right angles tend to make

geometry easier, don't break up the right angles. Instead, draw the segment $\overline{OB}$ to form the two

triangles. Since segments AB and CB are tangent to the circle, we know that segment OB bisects

$\angle ABC$, so m$\angle BOC =$ m$\angle AOB = 60°$. Look at one triangle at a time. Start with triangle BOC.

Since $\angle OCB$ is a right angle and m$\angle BOC = 60°$, m$\angle OBC = 30°$, making the triangle a 30-60-

90. The sides in a 30-60-90 triangle are in a ratio of $1{:}\sqrt{3}{:}2$. Since radius $\overline{OC}$ is opposite the 30°

and has a length of 1 (given), side $\overline{BC}$, which is opposite the 60° angle, has a length of $\sqrt{3}$, and

$\overline{OB}$, which is opposite the 90° angle, has a length of 2. Use $\overline{CB}$ as the base and $\overline{OC}$ as the height,

and plug these values into the area formula to get $A = \dfrac{1}{2}bh = \dfrac{1}{2}(\sqrt{3})(1) = \dfrac{\sqrt{3}}{2}$. Now look at

triangle AOB. Follow a similar line of reasoning to determine that this is also a 30-60-90 triangle

with sides 1, $\sqrt{3}$, and 2. (Alternatively, prove that the triangles are congruent using AAS or ASA.)

Since the triangles are congruent, double the area of the first triangle to get the total area:

$\dfrac{\sqrt{3}}{2} \times 2 = \sqrt{3}$.

REFLECT

Congratulations on completing Chapter 7!
Here's what we just covered.
Rate your confidence in your ability to:

- Understand tangent lines, secant lines, and chords, and the relationships among these and other parts of a circle

  ① ② ③ ④ ⑤

- Calculate the length of an arc, or the area of a sector of a circle

  ① ② ③ ④ ⑤

- Solve problems using the ratios of arc length : circumference, sector area : total area, and central angle : total angle

  ① ② ③ ④ ⑤

- Understand the relationships between inscribed and central angles

  ① ② ③ ④ ⑤

- Solve problems with concentric circles

  ① ② ③ ④ ⑤

- Understand radians, and convert between radians and degrees

  ① ② ③ ④ ⑤

If you rated any of these topics lower than you'd like, consider reviewing the corresponding lesson before moving on, especially if you found yourself unable to correctly answer one of the related end-of-chapter questions.

Access your online student tools for a handy, printable list of Key Points for this chapter. These can be helpful for retaining what you've learned as you continue to explore these topics.

Chapter 8
Circles: Constructions and Equations

8

GOALS By the end of this chapter, you will be able to:

- Construct the incenter, incircle, circumcenter, and circumcircle for triangles and polygons

- Construct the centroid and orthocenter for triangles

- Understand the standard equation of a circle and use the equation to solve for the center, radius, or coordinates on a circle

- Use the "completing the square" technique for a circle equation in expanded form

- Graph a circle from its equation

- Apply your understanding to general theorems and proofs

Lesson 8.1
Constructions

Supplies
You should have your compass and straightedge ready for this lesson.

In this lesson, you will learn how to construct **inscribed** and **circumscribed** circles on triangles and quadrilaterals.

CIRCUMCENTER/CIRCUMCIRCLE

This lesson will use several of the basic constructions that you learned in Lesson 2.3 of this book. You may find it helpful to refer to that lesson as needed.

All triangles can have circumscribed circles (also known as **circumcircles**). The **circumcenter** of a triangle is the intersection of the **perpendicular bisectors** of its sides.

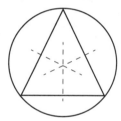  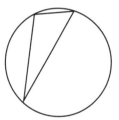

To find the circumcenter, you'll construct the perpendicular bisectors and see where they meet.

> When three or more lines intersect, the intersection point is known as a **point of concurrency**.

Then, to construct the circumcircle, you'll center the circle at the constructed circumcenter, and make the circle touch the triangle's vertices.

Here's how to do it.

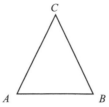

Supplies

Access your student tools to download larger, printable versions of the images in this section.

Construct the circumscribed circle of triangle *ABC*.

To find the circumcenter of the triangle, construct the triangle's perpendicular bisectors. It will be sufficient to construct just two of the three bisectors, since that will be enough to show the point of concurrency.

1. Construct the perpendicular bisector for side *AB*.

With the compass needle on vertex *A* and the drawing point on *B*, make a circle.

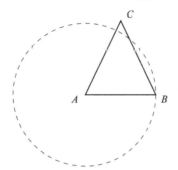

Then, with the compass needle on *B* and the drawing point on *A*, make another circle, which will intersect the first circle in two places.

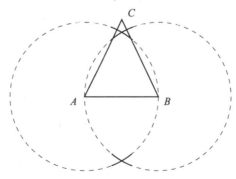

Next, use a straightedge to connect the two points where the arcs intersect, as shown. This new line is the perpendicular bisector for *AB*.

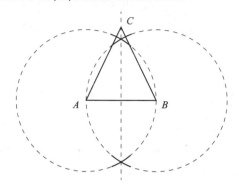

2. Construct the perpendicular bisector for side *BC*.

Repeat the same process to construct the perpendicular bisector for side *BC*. You may find it helpful to erase some of the previous arcs so that you can better see what you're doing.

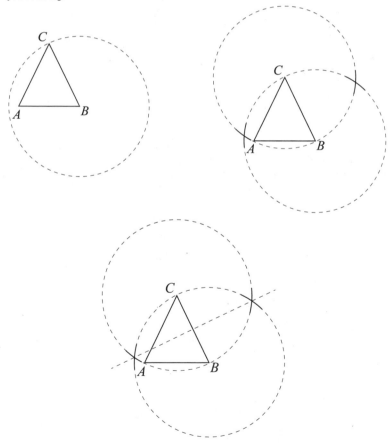

We have found the circumcenter of triangle *ABC*. For this exercise, we'll label the circumcenter as *O*.

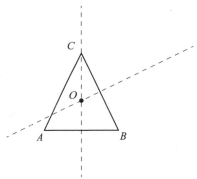

3. Draw the circumcircle.

Next, make a circle that touches the vertices of the triangle. Place the compass needle on point *O* and the drawing point on one of the vertices. Make a circle. If you did it correctly, you'll find that the circle touches each of the triangle's vertices.

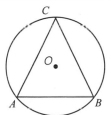

This method works to construct a circumcircle for *any* triangle. Try it on the ones below! Then, try a few more on your own for practice.

Special Case: Right Triangle

In a right triangle, the circumcenter will always be the midpoint of the hypotenuse. You can construct the circumcenter more easily by bisecting the hypotenuse.

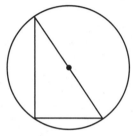

INCENTER/INCIRCLE

All triangles can have inscribed circles (also known as **incircles**). The **incenter** of a triangle is the point where the **angle bisectors** intersect. To find the incenter, you'll construct the angle bisectors of the triangle and see where they meet.

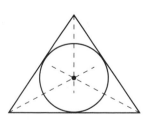 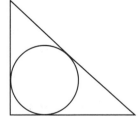

The incircle will be **tangent** to the triangle's sides. To find one of these points of tangency, you'll construct a perpendicular line that passes through the incenter.

Here's how to do it.

EXAMPLE 2

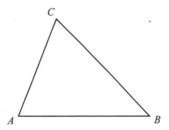

Construct the inscribed circle of triangle *ABC*.

To find the incenter of the triangle, construct the triangle's angle bisectors. We only need to construct two of the three angle bisectors, since that will be enough to show the point of concurrency.

1. Construct the angle bisector for angle *A*.

With the compass needle on vertex *A*, make an arc that passes through the two legs of that angle. For the purposes of this exercise, we'll call these two intersection points *M* and *N*.

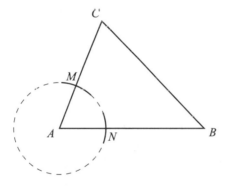

Now, position the compass needle point on *M* and the drawing point on *A*. With this radius, make a circle.

Repeat for point *N*, making a circle of the same radius, which will intersect the previous circle.

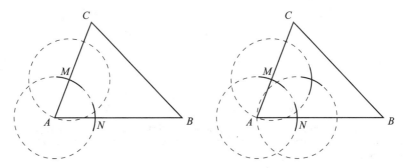

Finally, use a straightedge to draw a line through this new intersection and point *A*.

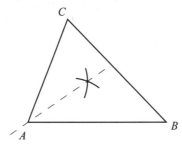

2. Construct the angle bisector for angle *B*.

Repeat the same process to construct the angle bisector for angle *B*. You may find it helpful to erase some of the previous arcs so that you can better see what you're doing.

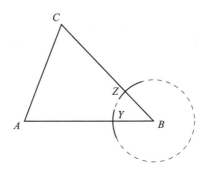

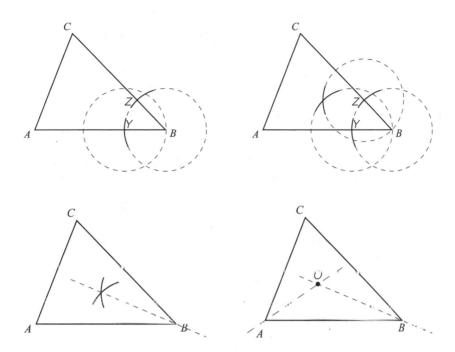

We have found the incenter of triangle *ABC*. For this exercise, we'll label the incenter as *O*.

Next, make a circle that is tangent to the sides of the triangle. In order to make sure that it's tangent, construct a line perpendicular to one of the sides. You only need to do this once; this will correctly identify the radius of our circle.

3. Construct a perpendicular line through the incenter.

With the compass needle on the incenter *O*, make an arc that intersects side *AB* in two places. For this exercise, we'll call these intersection points *P* and *Q*.

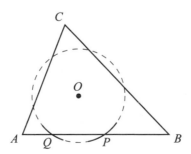

Now, with the compass needle point on *P* and the drawing point on *O*, make a circle.

Repeat for point *Q*, making a circle of the same radius, which will intersect the previous circle.

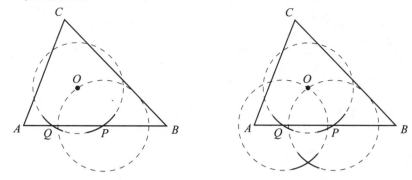

Use the straightedge to draw a line through this new intersection and point *O*.

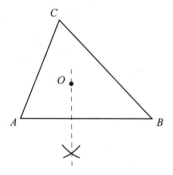

4. Draw the incircle.

This perpendicular intersection between the new line and side *AB* is one of the three tangent points for our circle. Now, the circle's center and radius are defined, and the circle can be drawn.

Position the compass needle at point *O*, and the drawing point at the perpendicular intersection that you have just drawn. Finally, make a circle. If you did it correctly, you'll find that the circle is tangent to each of the three sides of the triangle.

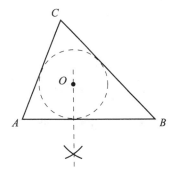

This method works to construct an incircle for *any* triangle. Try it on the ones below! Then, try a few more on your own for practice.

THE CENTROID AND ORTHOCENTER

Other than the circumcenter and incenter, mathematicians throughout history have identified *many*, many different definitions of a circle's "center." Many of these definitions are obscure and rarely used, but there are two that are quite common: the **centroid** and **orthocenter**.

Centroid

The centroid is the intersection of the **medians** of a triangle. A median is a segment from the triangle's vertex to the midpoint of the opposite side.

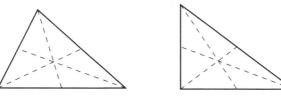

A fun fact about the centroid is that it is known as the triangle's "center of mass." In other words, if you had a physical triangle made out of wood or some other material, you could actually balance the triangle on a pencil point that is placed at the centroid. Likewise, if you hung such a triangle from a string that is placed at the centroid, it would balance there as well.

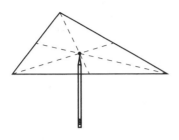
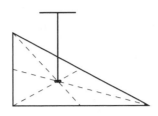

To construct a median of a triangle, first bisect one of the triangle's sides. (See "Perpendicular Bisector" in Lesson 2.3.) Then, connect the midpoint to the opposite vertex. Construct the remaining two medians to find the centroid.

Try it on the triangles below!

Orthocenter

The orthocenter is the intersection of the **altitudes** of a triangle. An altitude is a segment from the triangle's vertex that is perpendicular to the opposite side. The orthocenter can be outside the triangle.

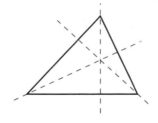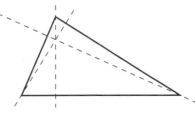

To construct an altitude of a triangle, construct a perpendicular line that passes through the opposite vertex. (See "Perpendicular Line Through a Point" in Lesson 2.3.) Construct two altitudes to find the orthocenter.

Try it on the triangles below!

Special Case: Equilateral Triangle

In an equilateral triangle, the incenter, circumcenter, centroid, and orthocenter are *all* found at the same point!

CIRCUMCENTER OF A QUADRILATERAL

As mentioned in Chapter 5, not all quadrilaterals have circumcircles. In fact, one of Euclid's discoveries was that a quadrilateral will have a circumcircle only if its opposite angles are **supplementary**. You won't always know the angles of a quadrilateral, so in order to construct a [possible] circumcircle, you may just go through a little trial and error.

If a quadrilateral (or other polygon) does have a circumcenter, then its definition is the same as that for a triangle—the circumcenter is the intersection of the perpendicular bisectors of the polygon's sides. To attempt construction of a polygon's circumcircle, construct each of the perpendicular bisectors of its sides. If and only if the perpendicular bisectors intersect at a single point, then the circumcircle can be constructed.

EXAMPLE 3

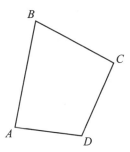

Construct the circumcircle for quadrilateral *ABCD*.

First, we can tell you that this quadrilateral does in fact have a circumcenter. To find it, construct the perpendicular bisectors of the quadrilateral's sides, and see where they intersect. For this exercise, we will assume that you are comfortable with constructing perpendicular bisectors. Again, refer to Lesson 2.3 if needed!

1. Construct the perpendicular bisectors for each side.

Construct the perpendicular bisector for *AB*:

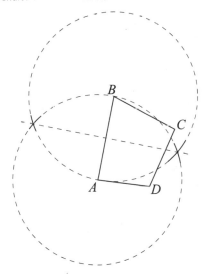

If you know that a quadrilateral has a circumcenter, then you only need to construct two perpendicular bisectors in order to find the circumcenter. It's more common that you wouldn't know whether or not the circumcenter can be constructed, so make a habit of constructing all four perpendicular bisectors.

Construct the perpendicular bisector for *BC*:

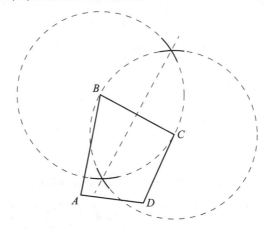

Construct the perpendicular bisector for *CD*:

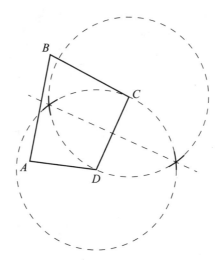

Construct the perpendicular bisector for *DA*:

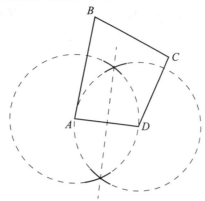

We have found the circumcenter of the quadrilateral.

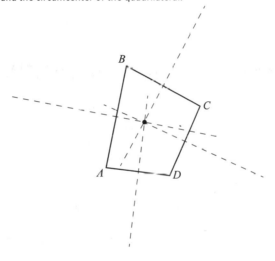

2. Draw the circumcircle.

Finally, construct the circle centered at the circumcenter and touching the vertices of the quadrilateral.

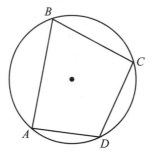

INCENTER OF A QUADRILATERAL

As mentioned in Chapter 5, not all quadrilaterals have incircles. One property of quadrilaterals that have incircles is that the two pairs of opposite sides have the same total length. You won't always know the side lengths of a quadrilateral, so in order to construct a [possible] incircle, you may just go through a little trial and error.

If a quadrilateral (or other polygon) does have an incircle, then its definition is the same as that for a triangle—the incenter is the intersection of the angle bisectors of the polygon. To attempt construction of a polygon's incircle, construct each of the angle bisectors. If and only if the angle bisectors intersect at a single point, then the incircle can be constructed.

EXAMPLE 4

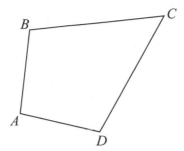

Construct the incircle for quadrilateral *ABCD*.

First, we can tell you that this quadrilateral does in fact have an incenter. To find it, construct the angle bisectors of the quadrilateral, and see where they intersect. For this exercise, we will assume that you are comfortable with constructing angle bisectors. Again, refer to Lesson 2.3 if needed!

1. Construct the angle bisectors for each angle.

Construct the angle bisector for *A*:

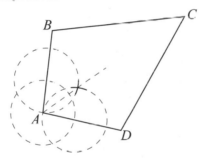

Construct the angle bisector for *B*:

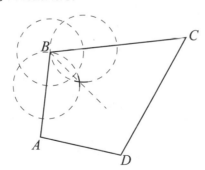

If you know that a quadrilateral has an incenter, then you only need to construct two angle bisectors in order to find the incenter. It's more common that you wouldn't know whether or not the incenter can be constructed, so make a habit of constructing all four angle bisectors.

Construct the angle bisector for *C*:

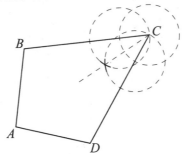

Construct the angle bisector for *D*:

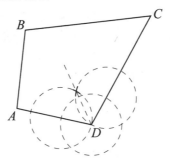

We have found the incenter of quadrilateral *ABCD*. For this exercise, we'll label the incenter as *O*.

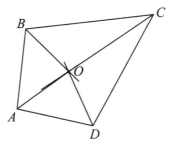

Next, make a circle that is tangent to the sides of the quadrilateral. In order to make sure that it's tangent, construct a perpendicular line to one of the sides. You need to do this only once; this will correctly identify the radius of the circle.

2. Construct a perpendicular line through the incenter.

With the compass needle on the incenter *O*, make an arc that intersects side *AB* in two places. For this exercise, we'll call these intersection points *P* and *Q*.

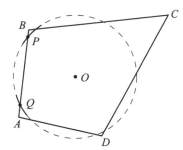

Now, position the compass needle point on *P* and the drawing point on *O*. With this radius, sweep the compass around and make an arc that's outside of the quadrilateral, on the opposite side of *AB*.

Repeat for point *Q*, making an arc of the same radius, which will intersect the previous arc.

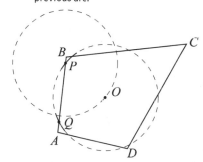 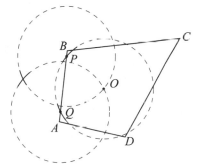

Use the straightedge to draw a line through this new intersection and point O.

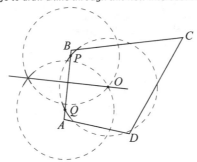

3. Draw the incircle.

This perpendicular intersection between the new line and side AB is one of the four tangent points for our circle. Now, the circle's center and radius are defined and the circle can be drawn.

Position the compass needle at point O, and the drawing point at the perpendicular intersection that you have just drawn. Finally, make a circle. If you did it correctly, you'll find that the circle is tangent to each of the four sides of the quadrilateral.

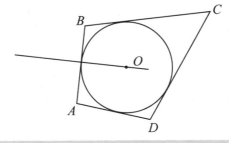

Lesson 8.2
Equation of a Circle

We know that the definition of a circle is "the set of points that are equidistant from a given center point." To express a circle as an equation, we can use the Pythagorean theorem or the distance formula.

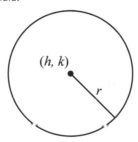

In this figure, the center of the circle is (h, k), and the radius is r. We know that for every point on the circle, the radius will always be equal to r. In other words, if (x, y) is any point on the circumference of the circle, then the distance between (x, y) and (h, k) will always be equal to r.

We know how to calculate distance using the Pythagorean theorem. If the distance from (x, y) to (h, k) were represented as a right triangle, like the one below, then we could say that $a^2 + b^2 = r^2$.

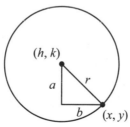

The values of a and b are the distances between the x- and y-coordinates, respectively. That is, the value of a is equal to $(x - h)$, and the value of b is equal to $(y - k)$.

$$a^2 + b^2 = r^2$$
$$(x - h)^2 + (y - k)^2 = r^2$$

That sure looks familiar! The equation of a circle is the same thing as the Distance Formula.

> **Equation of a Circle—Standard Form**
>
> For a circle with center (h, k) and radius r:
>
> $$(x - h)^2 + (y - k)^2 = r^2$$

And of course, if the center happens to be the origin $(0, 0)$, then the equation can be simplified:

> **Equation of a Circle**
>
> For a circle with center at the origin, and radius r:
>
> $$x^2 + y^2 = r^2$$

If you see an equation in the form of $(x - h)^2 + (y - k)^2 = r^2$, you'll know immediately that it is a circle with center (h, k) and radius r.

In the equation, the constants are h, k, and r, while x and y are variables representing all the possible points on the circle. Most commonly, you'll see the constants as numerical values and x and y as variables.

Find the equation for the circle with center (5, 4) and radius 6.

Substitute the given values into the standard equation for a circle:

$(x - h)^2 + (y - k)^2 = r^2$

$h = 5$
$k = 4$
$r = 6$

$(x - 5)^2 + (y - 4)^2 = 6^2$

EXAMPLE 6

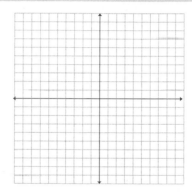

Graph the circle with the equation $(x - 2)^2 + (y + 3)^2 = 5^2$

From the equation, you know that the center is (2, –3), and the radius is 5. Begin by plotting the center at (2, –3).

Now, the easiest points to find would be parallel to the axes—that is, directly horizontal or vertical to the center. These points can be found by counting, and you don't need to use the equation at all.

The radius of the circle is 5. If you count from the center and move 5 units directly upward, you would find point (2, 2). This point will be on the circle, since it is exactly 5 units from the center.

Similarly, if you count 5 units down from the center, you would find point (2, –8).

5 units to the left of the center would be point (–3, –3).

5 units to the right of the center would be point (7, –3).

Plot these four points:

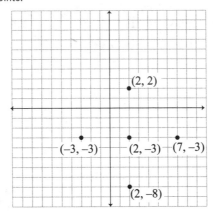

Then, connect the points with a smooth curve to make a circle. You could also use your compass.

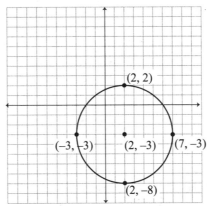

Fun fact: If there happens to be a Pythagorean triple that works with your radius, then you can also find eight more integer coordinates that lie on the circle rather easily. For example, the radius in this exercise is 5. Remember that 3-4-5 is a Pythagorean triple, or in other words, $3^2 + 4^2 = 5^2$. That means that we can add and subtract different combinations of 3 and 4 to the center coordinates to find more points on the circle.

For example, if you count from the center (2, −3) and move right 3, up 4, then you find point (5, 1). This point is on the circle because it has a distance of 5 from the center.

You can also move right 4, up 3, and find point (6, 0). This point is also 5 units from the center, and it lies on the circle.

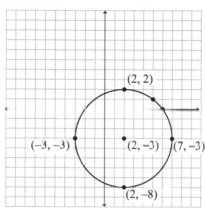

Here are all the integer coordinates that you can find in a similar manner for this circle:

$$(2 + 3, -3 + 4) = (5, 1)$$
$$(2 + 4, -3 + 3) = (6, 0)$$
$$(2 + 3, -3 - 4) = (5, -7)$$
$$(2 + 4, -3 - 3) = (6, -6)$$
$$(2 - 3, -3 + 4) = (-1, 1)$$
$$(2 - 4, -3 + 3) = (-2, 0)$$
$$(2 - 3, -3 - 4) = (-1, -7)$$
$$(2 - 4, -3 - 3) = (-2, -6)$$

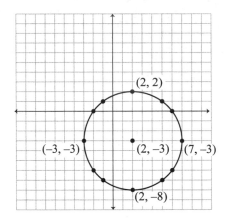

(2, 2)

(−3, −3) (2, −3) (7, −3)

(2, −8)

The rest of the coordinates won't be integer pairs. But you can find any other coordinate on the circle by choosing a value for x and solving for y, or vice versa, using the equation of a circle.

Sometimes, an equation will be given in expanded form, instead of the standard (factored) form $(x - h)^2 + (y - k)^2 = r^2$.

For example, the expanded form of $(x - h)^2$ is $x^2 - 2hx + h^2$, and the expanded form of $(y - k)^2$ is $y^2 - 2ky + k^2$.

So the full, expanded form of the equation might look like this:

$$x^2 - 2hx + h^2 + y^2 - 2ky + k^2 = r^2$$

The circle formula is $(x - h)^2 + (y - k)^2 - r^2$, where (h, k) is the center of the circle and r is the radius. Use Process of Elimination. Choices (F) and (G) can be eliminated because they do not square the radius. Choice (K) can also be eliminated because the x-coordinate of the center is +2, so the first part of the equation should be $(x - 2)^2$, not $(x + 2)^2$. Choice (H) can be eliminated because the y-coordinate is −6, so the second part should be $(y + 6)^2$ not $(y - 6)^2$. Only (J) remains. The correct answer is (J).

With the equation in this form (with various numbers substituted for the variables), it can be difficult to spot the constants h, k, and r. So to solve these equations, we'll factor the expanded form using a technique called "completing the square."

For example, if you have the equation $x^2 + 14x + 13 = 0$, it might seem like it can't be factored, since there's no numbers that add to 13 and multiply to 14. "Completing the square" means manipulating the equation so that it can be factored in the form of $(x + n)^2$.

Remember that $(x + a)^2$ will always expand to $x^2 + 2ax + a^2$. Therefore, the value of $2a$ will always match the coefficient of your middle term. (In this example, the coefficient is 14). So, if you divide the coefficient by 2, you find a.

Here's how to do it:

$x^2 + 14x + 13 = 0$
$x^2 + 14x = -13$ Subtract 13 from both sides of the equation.

We are left with a quadratic expression in the form of $x^2 + 14x +$ ___, and we need to fill in the blank. Half of 14 is 7, so the binomial factor we want is $(x + 7)$.

This expands as $(x + 7)^2 = x^2 + 14x + 49$.

So we're substituting $(x^2 + 14x)$ for $(x^2 + 14x + 49)$, which means we're adding 49 to the equation. Don't forget to add it to the right side as well!

$x^2 + 14x + \mathbf{49} = -13 + \mathbf{49}$ Add 49 to both sides of the equation.
$x^2 + 14x + 49 = 36$ Simplify.
$(x + 7)^2 = 36$ Substitute $x^2 + 14x + 49 = (x + 7)^2$.
$x + 7 = \pm 6$ Take the square root of both sides of the equation.

Therefore, the solutions to this equation are $x = -1$ and $x = -13$.

When you need to factor the equation of a circle, you're typically going to use "completing the square" twice in order to factor the x terms and the y terms separately.

Find the center and the radius of the circle with the equation $x^2 + y^2 - 10x - 12y = 3$.

$x^2 + y^2 - 10x - 12y = 3$

$x^2 - 10x + y^2 - 12y = 3$ Put the x terms and y terms next to each other.

Complete the square for the x terms:

$x^2 - 10x + 25 = (x - 5)^2$ The coefficient is -10; half of -10 is -5.

Complete the square for the y terms:

$y^2 - 12y + 36 = (y - 6)^2$ The coefficient is -12; half of -12 is -6.

Therefore, we're adding 25 and 36 to both sides of the equation.

$x^2 - 10x + \mathbf{25} + y^2 - 12y + \mathbf{36} = 3 + \mathbf{25} + \mathbf{36}$ Add $25 + 36$ to both sides of the equation.

$x^2 - 10x + 25 + y^2 - 12y + 36 = 64$ Simplify.

$(x - 5)^2 + y^2 - 12y + 36 = 64$ Substitute $x^2 - 10x + 25 = (x - 5)^2$.

$(x - 5)^2 + (y - 6)^2 = 64$ Substitute $y^2 - 12y + 36 = (y - 6)^2$.

$(x - 5)^2 + (y - 6)^2 = 8^2$ Substitute $64 = 8^2$.

The equation is now in standard form. Therefore, $h = 5$, $k = 6$ (so the center is $(5, 6)$), and $r = 8$.

Lesson 8.3
Transformations

In this section, you will apply what you've learned about circles to problems, theorems, and proofs.

EXAMPLE 8

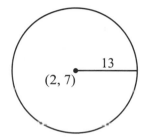

In the figure above, the circle is in the standard coordinate plane, with center (2, 7) and radius 13.

For each of the coordinates below, determine whether the point is inside the circle, outside the circle, or on the circle.

1. (−3, 19)	3. (16, 14)	5. (−10, 12)
2. (6, −5)	4. (−14, 2)	6. (5, 14)

Anytime you're asked to prove whether or not a point is on a circle, the question you're really being asked is distance. Your task is to determine whether or not the distance for each pair of coordinates is equal to the radius of the circle.

We know that the radius of the circle is 13. That means that if a coordinate is on the circle, then it is exactly 13 units from the center. What does it mean if the coordinate is less than 13 units from the center? More than 13?

Use the distance formula to find the distance between each point and the center of the circle (2, 7).

We'll do #1 step by step. Try the rest on your own, and then check your answers against ours.

1. $d = \sqrt{(x_1 - x_2)^2 + (y_1 - y_2)^2}$

$d = \sqrt{(2 - (-3))^2 + (7 - 19)^2}$ Substitute the coordinates (2, 7) and (–3, 19).

$d = \sqrt{(5)^2 + (-12)^2}$ Simplify.

$d = \sqrt{25 + 144}$

$d = \sqrt{169}$

$d = 13$

The distance is 13 (equal to the radius), so this point is **on the circle**.

2. $d = \sqrt{(2 - 6)^2 + (7 - (-5))^2}$
   $d \approx 12.65$

The distance is ≈ 12.65 (less than the radius), so this point is **inside the circle**.

3. $d = \sqrt{(2 - 16)^2 + (7 - 14)^2}$
   $d \approx 15.65$

The distance is ≈ 15.65 (greater than the radius), so this point is **outside the circle**.

4. $d = \sqrt{(2 - (-14))^2 + (7 - 2)^2}$
   $d \approx 16.76$

The distance is ≈ 16.76 (greater than the radius), so this point is **outside the circle**.

Here is how you may see the equation of a circle on the SAT.

The standard form of the equation of a circle is $(x - h)^2 + (y - k)^2 = r^2$, where the center of the circle is at point (h, k) and the radius of the circle is r. What is the standard form of the equation of the circle defined by the equation $x^2 + y^2 - 6x + 8y = 0$?

   A. $(x - 6)^2 + (y + 8)^2 = 0$
   B. $(x + 3)^2 + (y + 4)^2 = 25$
   C. $(x - 3)^2 + (y + 4)^2 = 25$
   D. $(x - 3)^2 + (y + 4)^2 = 5$

5. $d = \sqrt{(2-(-1\,0)^2 + (7-12\,)^2}$

   $d = 13$

The distance is 13 (equal to the radius), so this point is **on the circle**.

6. $d = \sqrt{(2-5)^2 + (7-1\,4)^2}$

   $d \approx 7.62$

The distance is ≈ 7.62 (less than the radius), so this point is **inside the circle**.

EXAMPLE 9

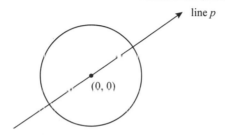

line p

$(0, 0)$

In the figure above, the circle has the equation $x^2 + y^2 = 5^2$. Line p passes through the center of the circle, and it has the equation $y = \left(\dfrac{4}{3}\right)x$.

Line q (not shown) has the equation $y = \left(-\dfrac{3}{4}\right)x + \left(\dfrac{25}{4}\right)$, and it passes through the point (3, 4). Is line q tangent to the circle?

Recall the definition of a tangent line: a line that intersects the circle at a single point, and is perpendicular to the radius at the point of tangency. In order to determine whether the line is tangent, we'll need to prove both of those things.

Perhaps an easy place to start is to compare the equations of the two lines. Since line p passes through the center of the circle, we know that a radius exists on that line. Therefore, it should be perpendicular to line q, if line q is tangent to the circle.

Recall that two lines are perpendicular if they have opposite reciprocal slopes.

The slopes of lines p and q are $\left(\dfrac{4}{3}\right)$ and $\left(-\dfrac{3}{4}\right)$, respectively. Are those opposite reciprocals? Yes! This is proof that the lines could be tangent.

We will also need to know if the lines intersect *at the point of tangency.* We could solve the two linear equations as a system. However, we were given a point for line q (3, 4). Why not try that one? Let's see if it satisfies the equation for line p.

$$y = \frac{4}{3}x$$

$$4 = \left(\frac{4}{3}\right) \times 3 \qquad \text{Substitute (3, 4) for } (x, y).$$

$$4 = 4 \qquad \qquad \text{✔ True!}$$

This means that the two lines intersect at point (3, 4).

Now, try the point (3, 4) in the equation for the circle. If the point lies on the circle, that means that it is an intersection point for *both* lines as well as the circle.

$$x^2 + y^2 = 5^2$$

$$3^2 + 4^2 = 5^2 \qquad \text{Substitute (3, 4) for } (x, y).$$

$$9 + 16 = 25 \qquad \text{✔ True!}$$

To recap, we now know that the point (3, 4) is the intersection of both lines and the circle. We also know that lines p and q are perpendicular to each other. Therefore, we have proved that line q is tangent to the circle.

Start by rewriting the equation with the x terms and y terms listed together:

$$x^2 - 6x + y^2 + 8y = 0$$

Next, to get the equation into standard form, you want to complete the square. Start with the x terms. Take half of the coefficient on the x (–6), square it, and add that to both sides. Half of –6 is 3, and 3^2 is 9, so add 9 to both sides of the equation:

$$x^2 - 6x + 9 + y^2 + 8y = 9$$

Now $x^2 - 6x + 9$ is a perfect square; it factors into $(x - 3)^2$, so you can rewrite the equation:

$$(x - 3)^2 + y^2 + 8y = 9$$

You can do the same to the y terms. Half of 8 is 4, and $4^2 = 16$, so add 16 to both sides of the equation:

$$(x - 3)^2 + y^2 + 8y + 16 = 9 + 16$$

Now $y^2 + 8y + 16$ is a perfect square; it factors into $(x + 4)^2$, so you can rewrite the equation again:

$$(x - 3)^2 + (y + 4)^2 = 25$$

That matches (D). Note you do not need to take the square root of 25; that would change the value of the equation. Instead, the radius is left squared. The correct answer is (C).

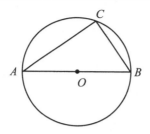

Given: *AB* is a diameter of the circle.
C is a point on the circle.
O is the center of the circle.

Prove: ∠*ACB* is 90°

One way to prove that ∠*CAB* is 90° is to use the fact that all radii are equal. Since *C* is a point on the circle, we know that *OC* is a radius. That means that *OC* is congruent to *OA* and *OB*.

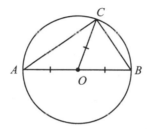

Because the two radii are equal, we know that the two smaller triangles are both isosceles. That means that each one has a pair of congruent base angles.

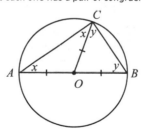

We also know that the angles in triangle *ABC* must have a sum of 180°. Therefore, we can set up an equation:

$$x + y + x + y = 180°$$

$$2x + 2y = 180° \qquad \text{Simplify.}$$

$$2(x + y) = 180°$$

$$x + y = \frac{180°}{2}$$

$$x + y = 90°$$

We have proved that the angle *ACB*, which is equal to *x* + *y*, must be equal to 90°.

EXAMPLE 🔒 11

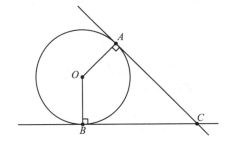

Given: *O* is the center of the circle.
$\overline{AC}$ is tangent to the circle at point *A*.
$\overline{BC}$ is tangent to the circle at point *B*.

Prove: $\overline{AC} \cong \overline{BC}$

One way to prove that $\overline{AC} \cong \overline{BC}$ is by using congruent triangles. If we split the quadrilateral down the middle, we'd have two right triangles, as shown:

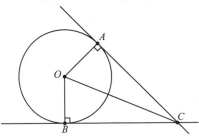

We know that $\overline{OA} \cong \overline{OB}$, because they are both radii. Also, we know of course that $\overline{OC} \cong \overline{OC}$, by the Reflexive Property.

Therefore, we have two right triangles with two pairs of congruent, corresponding sides. The triangles must be congruent. You can refer to the **Hypotenuse-Leg Theorem**, which states that if two right triangles have a hypotenuse and leg pair that are congruent, then the two triangles are congruent.

Finally, if the two triangles are congruent, then we know that the corresponding sides $\overline{AC}$ and $\overline{BC}$ must be congruent.

CHAPTER 8 PRACTICE QUESTIONS

Directions: Complete the following problems as specified by each question. For extra practice after answering each question, try using an alternative method to solve the problem or check your work. Larger, printable versions of images are available in your online student tools.

1. Using a compass and a straightedge, construct an equilateral triangle with side *AB*.

3. Using a compass and a straightedge, construct the incircle of the regular hexagon shown below.

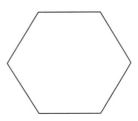

2. Using a compass and a straightedge, construct the circumcircle of right triangle *ABC* below.

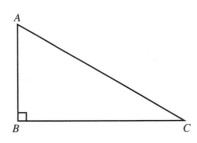

4. Is the point (–3, –2) inside, outside, or on the circle $x^2 + y^2 - 6x + y^2 + 8y = 39$?

5. Graph the circle represented by the equation $x^2 + y^2 + 4x - 2y = 20$.

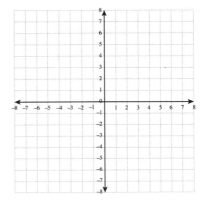

6. In the figure below, $\overline{XY}$ and $\overline{XZ}$ are tangent to the circle with center O at Y and Z, respectively, and $\angle YXZ$ is a right angle.

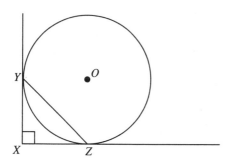

Is $\triangle XYZ$ isosceles? Why?

7. In the figure below, diameters $\overline{AC}$ and $\overline{BD}$ intersect at point O, and O is the center of the circle.

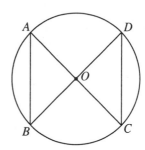

Is $\overline{AB} \cong \overline{CD}$? Why?

8. In the figure below, AB is tangent to circles P and Q at points A and B, respectively, and DC is tangent to circles P and Q at points D and C, respectively. $\overline{AD}$ is a diameter of circle P and $\overline{BC}$ is a diameter of circle Q.

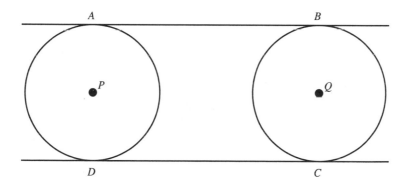

Are circles P and Q are congruent? Why?

SOLUTIONS TO CHAPTER 8 PRACTICE QUESTIONS

1. An equilateral triangle has three congruent sides. To form an equilateral triangle, find a point (call it C for reference) such that $\overline{AB} \cong \overline{BC} \cong \overline{CA}$. To make sure that $\overline{AB} \cong \overline{CA}$, construct a circle with center A and radius $\overline{AB}$.

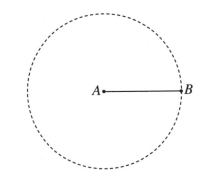

Any other radius of this circle is congruent to $\overline{AB}$. To make sure that $\overline{AB} \cong \overline{BC}$, construct a circle with center A and radius $\overline{AB}$.

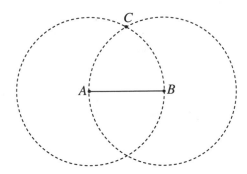

Any other radius of this circle is congruent to $\overline{AB}$. There are two points that are on both circles. Use the upper point. Call that C. $\overline{AB}$ is a radius of both circle A and circle B, so it is congruent to both $\overline{AC}$ and $\overline{CB}$, so triangle ABC is equilateral.

The same would be true if you used the bottom intersection point as the third vertex.

2. To draw a circumcircle of a triangle, first find the circumcenter. In the case of a right triangle, there is a shortcut: The circumcenter is always the midpoint of the hypotenuse. To find the midpoint of a segment, construct the perpendicular bisector. To do this use a compass to construct a circle with needle point on *A* and drawing point on *C*. Now reverse this. Construct a circle with needle point on *C* and drawing point on *A*. The circles have two points of intersection. Use a straight edge to draw a line to connect the points of intersection.

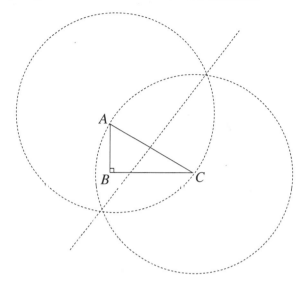

Find the point that the bisector intersects the hypotenuse. That is the midpoint. Place the needle point of the compass here and the drawing point on any of the three vertices of the triangle. Construct this circle.

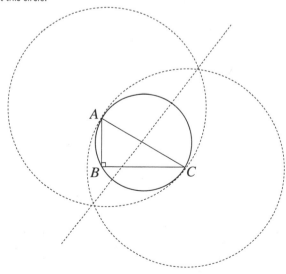

3. A polygon has an incircle if opposite sides are congruent. Since the hexagon is a regular hexagon, all sides (and thus opposite sides) are congruent, so this figure does have an incircle. To find the incircle, find the incenter, which is the intersection of the figure's angle bisectors. Pick two points. As an example, use the top two vertices on the hexagon. Start with the top left. Construct a circle with the compass needle on the top left vertex through the two adjacent sides. Construct congruent circles with the compass needle on the points of intersection. Use a straight edge to construct a line through the vertex and the point of intersection between the second two circles.

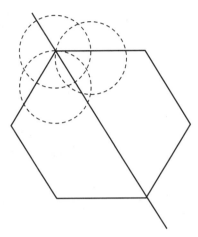

Now do the same for the top right vertex.

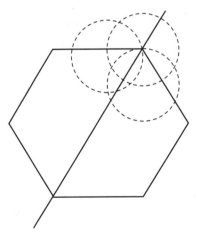

The point of intersection of these two lines is the incenter. An incircle is tangent to any of the sides of the hexagon. To find the point of tangency, construct a perpendicular line from the incenter to any side. Use the compass to draw a circle through two points on one side of the circle. For example, use the bottom side. Now, construct two circles, one with the drawing point at the incenter and the needle point at one intersection point and another with the drawing point at the incenter and the needle point at the other intersection point. Draw a line through the incenter and the other intersection point on these two circles. It crosses the side of the hexagon at the tangent point.

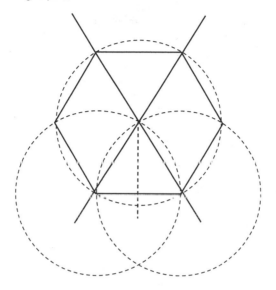

Finally, with the needle point at the incenter and the drawing point at the tangent point, construct an incircle of the hexagon.

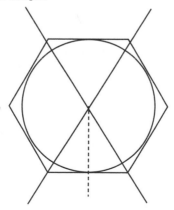

4. **Inside.**
Consider the definition of a circle: the set of points that is a given distance from a given point. That distance is called the radius, and that point is called the center. If a point is a distance equal to the radius from the center, the point is on the circle. If a point is a distance less than the radius from the center, the point is inside the circle. If a point is a distance greater than the radius from the center, the point is outside the circle. Determine the center and the radius of this circle. In the standard form of the equation, $(x - h)^2 + (y - k)^2 = r^2$, the center is (h, k) and the radius is r. The equation is given in expanded form, so put it into standard form using the complete the square method. Get the x terms together and the y terms together to get $(x^2 - 6x) + (y^2 + 8y) = 39$. Add the square of half the coefficient on x to both sides. The coefficient on x is −6, half of −6 is −3, and the square of −3 is 9, so add 9 to both sides to get $(x^2 - 6x + 9) + (y^2 + 8y) = 48$. Add the square of half the coefficient on y to both sides. The coefficient on y is 8, half of 8 is 4, and the square of 4 is 16, so add 16 to both sides to get $(x^2 - 6x + 9) + (y^2 + 8y + 16) = 64$. Factor the expression in each set of parentheses to get $(x - 3)^2 + (y + 4)^2 = 8^2$. Therefore, the center is (3, −4) and the radius is 8.

Next, determine how far the point (−3, −2) is from the point (3, −4). Use the distance formula: $d = \sqrt{(x_2 - x_1)^2 + (y_2 - y_1)^2} = \sqrt{(-3 - 3)^2 + (-2 - (-4))^2} = \sqrt{6^2 + 2^2} = \sqrt{40}$. Since $36 < 40 < 49$, $6 < \sqrt{40} < 7$. Therefore, the distance between (−3, −2) and the center is less than the length of the radius, so (−3, −2) is inside the circle.

5. To graph a circle, get its center and radius. To do this, get the equation of the circle into standard form: $(x - h)^2 + (y - k)^2 = r^2$ with center (h, k) and radius r. The equation is given in expanded form, so complete the square. First, group the x terms and the y terms: $(x^2 + 4x) + (y^2 - 2y) = 20$. Now, add the square of half the coefficient on x to both sides. The coefficient on x is 4, half is 2, and the square of 2 is 4, so add 4 to both sides to get $(x^2 + 4x + 4) + (y^2 - 2y) = 24$. Now, add the square of half the coefficient on y to both sides. The coefficient on y is −2, half is −1, and the square of −1 is 1, so add 1 to both sides to get $(x^2 + 4x + 4) + (y^2 - 2y + 1) = 25$. Factor the two expressions in parentheses to get $(x + 2)^2 + (y - 1)^2 = 5^2$. Therefore, the center of the circle is (−2, 1) and the radius is 5. Draw the center of the circle at (−2, 1) and draw points on the circle that are 5 away from the center.

Take advantage of the fact that 5 is the hypotenuse of a 3:4:5 right triangle. One point on the circle is 3 units to the right of and 4 units above the center: (−2 + 3, 1 + 4) = (1, 5). And so on: You can also find the points (1, −3), (−5, −3), (−5, 5), (2, 4), (2, −2), (−6, −2), and (−6, 4).

Also, plot the points 5 units above, 5 units below, 5 units to the left, and 5 units to the right of center: (–2, 6), (–2, –4), (–7, 1), and (3, 1). Plot these points on the graph and draw a smooth curve to connect them.

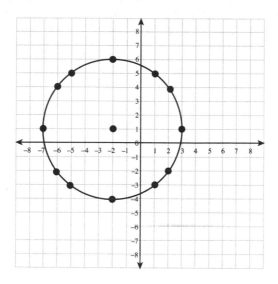

6. **Yes.**

The goal is to prove that ΔXYZ is isosceles. Notice that ΔXYZ is a right triangle. The hypotenuse of a right triangle cannot be congruent to another side in the triangle. Therefore, to show that it is isosceles, prove that $\overline{YX} \cong \overline{XZ}$. The question says that $\overline{YX}$ is a tangent to the circle at Y, so draw a radius to Y and mark it as a right angle. The question also says that $\overline{XZ}$ is a tangent to the circle at Z, so draw a radius to Z and mark it as a right angle. This forms quadrilateral OXYZ, which has three right angles. Since the sum of the angles in a quadrilateral is 360, the fourth angle must also be a right angle, making the figure a rectangle. Since $\overline{OZ}$ and $\overline{OY}$ are both radii, they are congruent. If adjacent sides of a rectangle are congruent, the rectangle is a square. If the rectangle is a square, then $\overline{YX} \cong \overline{XZ}$, which, as described above, proves that ∠XYZ is a 45° angle.

7. **Yes.**

The goal is to prove that $\overline{AB} \cong \overline{CD}$. Each of these segments is part of a triangle. When this is the case, the method will often be to prove that the two triangles are congruent. Try to find a way to use one of the congruence postulates/theorems: SSS, SAS, ASA, or AAS. Find congruent corresponding sides and angles. The center of the circle is O, so $\overline{AO}$, $\overline{BO}$, $\overline{CO}$, and $\overline{DO}$ are all radii and thus congruent. Since $\overline{AC}$ and $\overline{BD}$ intersect at O, ∠AOB and ∠DOC are vertical angles and thus are congruent. Therefore, $\overline{AO} \cong \overline{DO}$, $\overline{BO} \cong \overline{CO}$, and ∠AOB ≅ ∠DOC. Thus, the SAS postulate can be used to prove that triangles AOB and DOC are congruent. Thus, $\overline{AB} \cong \overline{CD}$.

8. **Yes.**

The information in the question says that $\overline{AB}$ is tangent to circles P and Q at points A and B, respectively, and that $\overline{DC}$ is tangent to circles P and Q at points D and C, respectively. Since tangents, by definition, are perpendicular to the radii at the point of intersection, draw the four radii and mark them as perpendicular. Notice that $\overline{AP}$ and $\overline{PD}$ form diameter $\overline{AD}$ and that $\overline{BQ}$ and $\overline{QC}$ form diameter $\overline{BC}$. Notice further that these segments form quadrilateral $ABCD$ and that this quadrilateral has four right angles. By definition, a quadrilateral with four right angles is a rectangle, so $ABCD$ is a rectangle. The goal is to prove that the two circles are congruent. This can be done by showing that they have congruent radii. In this case, the diameters are opposite sides in a rectangle. By a property of a rectangle, opposite sides are congruent. Since the diameters are congruent, the radii must be congruent and so are the circles.

REFLECT

**Congratulations on completing Chapter 8!
Here's what we just covered.
Rate your confidence in your ability to:**

- Construct the incenter, incircle, circumcenter, and circumcircle for triangles and polygons

  ① ② ③ ④ ⑤

- Construct the centroid and orthocenter for triangles

  ① ② ③ ④ ⑤

- Understand the standard equation of a circle and use the equation to solve for the center, radius, or coordinates on a circle

  ① ② ③ ④ ⑤

- Use the "completing the square" technique for a circle equation in expanded form

  ① ② ③ ④ ⑤

- Graph a circle from its equation

  ① ② ③ ④ ⑤

- Apply your understanding to general theorems and proofs

  ① ② ③ ④ ⑤

If you rated any of these topics lower than you'd like, consider reviewing the corresponding lesson, especially if you found yourself unable to correctly answer one of the related end-of-chapter questions.

Access your online student tools for a handy, printable list of Key Points for this chapter. These can be helpful for retaining what you've learned as you continue to explore these topics.

NOTES

NOTES

International Offices Listing

China (Beijing)
1501 Building A,
Disanji Creative Zone,
No.66 West Section of North 4th Ring Road Beijing
Tel: +86-10-62684481/2/3
Email: tprkor01@chol.com
Website: www.tprbeijing.com

China (Shanghai)
1010 Kaixuan Road
Building B, 5/F
Changning District, Shanghai, China 200052
Sara Beattie, Owner: Email: sbeattie@sarabeattie.com
Tel: +86-21-5108-2798
Fax: +86-21-6386-1039
Website: www.princetonreviewshanghai.com

Hong Kong
5th Floor, Yardley Commercial Building
1-6 Connaught Road West, Sheung Wan, Hong Kong
(MTR Exit C)
Sara Beattie, Owner: Email: sbeattie@sarabeattie.com
Tel: +852-2507-9380
Fax: +852-2827-4630
Website: www.princetonreviewhk.com

India (Mumbai)
Score Plus Academy
Office No.15, Fifth Floor
Manek Mahal 90
Veer Nariman Road
Next to Hotel Ambassador
Churchgate, Mumbai 400020
Maharashtra, India
Ritu Kalwani: Email: director@score-plus.com
Tel: + 91 22 22846801 / 39 / 41
Website: www.score-plus.com

India (New Delhi)
South Extension
K-16, Upper Ground Floor
South Extension Part–1,
New Delhi-110049
Aradhana Mahna: aradhana@manyagroup.com
Monisha Banerjee: monisha@manyagroup.com
Ruchi Tomar: ruchi.tomar@manyagroup.com
Rishi Josan: Rishi.josan@manyagroup.com
Vishal Goswamy: vishal.goswamy@manyagroup.com
Tel: +91-11-64501603/ 4, +91-11-65028379
Website: www.manyagroup.com

Lebanon
463 Bliss Street
AlFarra Building - 2nd floor
Ras Beirut
Beirut, Lebanon
Hassan Coudsi: Email: hassan.coudsi@review.com
Tel: +961-1-367-688
Website: www.princetonreviewlebanon.com

Korea
945-25 Young Shin Building
25 Daechi-Dong, Kangnam-gu
Seoul, Korea 135-280
Yong-Hoon Lee: Email: TPRKor01@chollian.net
In-Woo Kim: Email: iwkim@tpr.co.kr
Tel: + 82-2-554-7762
Fax: +82-2-453-9466
Website: www.tpr.co.kr

Kuwait
ScorePlus Learning Center
Salmiyah Block 3, Street 2 Building 14
Post Box: 559, Zip 1306, Safat, Kuwait
Email: infokuwait@score-plus.com
Tel: +965-25-75-48-02 / 8
Fax: +965-25-75-46-02
Website: www.scorepluseducation.com

Malaysia
Sara Beattie MDC Sdn Bhd
Suites 18E & 18F
18th Floor
Gurney Tower, Persiaran Gurney
Penang, Malaysia
Email: tprkl.my@sarabeattie.com
Sara Beattie, Owner: Email: sbeattie@sarabeattie.com
Tel: +604-2104 333
Fax: +604-2104 330
Website: www.princetonreviewKL.com

Mexico
TPR México
Guanajuato No. 242 Piso 1 Interior 1
Col. Roma Norte
México D.F., C.P.06700
registro@princetonreviewmexico.com
Tel: +52-55-5255-4495
+52-55-5255-4440
+52-55-5255-4442
Website: www.princetonreviewmexico.com

Qatar
Score Plus
Office No: 1A, Al Kuwari (Damas)
Building near Merweb Hotel, Al Saad
Post Box: 2408, Doha, Qatar
Email: infoqatar@score-plus.com
Tel: +974 44 36 8580, +974 526 5032
Fax: +974 44 13 1995
Website: www.scorepluseducation.com

Taiwan
The Princeton Review Taiwan
2F, 169 Zhong Xiao East Road, Section 4
Taipei, Taiwan 10690
Lisa Bartle: Email: lbartle@princetonreview.com.tw
Tel: +886-2-2751-1293
Fax: +886-2-2776-3201
Website: www.PrincetonReview.com.tw

Thailand
The Princeton Review Thailand
Sathorn Nakorn Tower, 28th floor
100 North Sathorn Road
Bangkok, Thailand 10500
Thavida Bijayendrayodhin (Chairman)
Email: thavida@princetonreviewthailand.com
Mitsara Bijayendrayodhin (Managing Director)
Email: mitsara@princetonreviewthailand.com
Tel: +662-636-6770
Fax: +662-636-6776
Website: www.princetonreviewthailand.com

Turkey
Yeni Sülün Sokak No. 28
Levent, Istanbul, 34330, Turkey
Nuri Ozgur: nuri@tprturkey.com
Rona Ozgur: rona@tprturkey.com
Iren Ozgur: iren@tprturkey.com
Tel: +90-212-324-4747
Fax: +90-212-324-3347
Website: www.tprturkey.com

UAE
Emirates Score Plus
Office No: 506, Fifth Floor
Sultan Business Center
Near Lamcy Plaza, 21 Oud Metha Road
Post Box: 44098, Dubai
United Arab Emirates
Hukumat Kalwani: skoreplus@gmail.com
Ritu Kalwani: director@score-plus.com
Email: info@score-plus.com
Tel: +971-4-334-0004
Fax: +971-4-334-0222
Website: www.princetonreviewuae.com

Our International Partners

The Princeton Review also runs courses with a variety of
partners in Africa, Asia, Europe, and South America.

Georgia
LEAF American-Georgian Education Center
www.leaf.ge

Mongolia
English Academy of Mongolia
www.nyescm.org

Nigeria
The Know Place
www.knowplace.com.ng

Panama
Academia Interamericana de Panama
http://aip.edu.pa/

Switzerland
Institut Le Rosey
http://www.rosey.ch/

All other inquiries, please email us at
internationalsupport@review.com